PAPERMAKING FIBERS

PAPERMAKING FIBERS

A Photomicrographic Atlas

Edited by WILFRED A. CÔTÉ

Co-published by the

RENEWABLE MATERIALS INSTITUTE of the

STATE UNIVERSITY of NEW YORK

COLLEGE of ENVIRONMENTAL SCIENCE and FORESTRY and

SYRACUSE UNIVERSITY PRESS

1980

This is Publication No. 1 in the Renewable Materials Institute Series. Its publication was supported by the College of Forestry Foundation, Inc. of the State University of New York College of Environmental Science and Forestry.

Manufactured in the United States of America

PREFACE

The original version of this book appeared in 1931 under the title *Atlas of Papermaking Fibers.* It was published by the New York State College of Forestry as Technical Publication No. 35 and was prepared by Charles H. Carpenter. In 1952 a revised and enlarged version appeared as Technical Publication No. 74. Professor Lawrence Leney added many new fibers, a new format was designed, and an improved method of reproduction was employed. *91 Papermaking Fibers* found wide use in the pulp and paper field as well as in teaching students, most of whom were majoring in paper science.

Following essentially the same design as the 1952 version, Technical Publication No. 74 was revised by Dr. Harold A. Core, Mr. Arnold C. Day, and Dr. Wilfred A. Côté in 1963. Some new species were added in the revision, and cross-sections of some man-made fibers that might be encountered in paper manufacture were also included. The 1963 revision reflected some changes in nomenclature to conform with the *Checklist of United States Trees**. The constructive criticisms of Professor Leney and the concurrence of Mr. Carpenter were helpful in the completion of the publication whose title was changed to *Papermaking Fibers.*

Because of the continuing demand for a handbook of this type, it was decided to produce yet another version when the 1963 publication went out of print. This work was started in 1978 and was a major undertaking because of the loss of all of the original plates by the contract printer following the 1963 publications. In spite of this handicap, and perhaps because of it, a new approach was made possible. A larger format was selected to provide for clearer detail in the photomicrographs. Better prints were prepared from the original negatives, in many instances. Also, the availability of scanning electron microscopy made it possible to add an introduction which none of the earlier versions included.

Among the contributors to this publication are Mr. J. J. McKeon, Mr. A. C. Day, Dr. R. B. Hanna, Ms. K. Kaufman, and Ms. J. Barton. Staff members of the College Publications Office who were of direct assistance include Mr. R. Karns, Mr. J. Novado, Ms. G. Owczarzak, Ms. F. Lawton, and Ms. C. Mudge. The project could

*Elbert L. Little, Jr. *Checklist of United States Trees (Native and Naturalized)*, U.S.D.A. Forest Service. Agriculture Handbook No. 541. 1979.

not have been completed without the willing cooperation of all the contributors.

Just as wood is variable in gross appearance, so too are the individual wood cell types variable in size, shape, and distinctive markings. Fortunately, for a particular kind of woody fiber, these departures are within limits that permit an experienced observer to make accurate identifications. It would be impossible to show all of the variations that may be present in a species in a few photomicrographs. However, in the preparation of this and earlier versions, an attempt was made to select features and characteristics that were considered typical. It should be emphasized that a single feature rarely serves for positive identification, but that the sum of the characteristics should be used.

Softwood and hardwood fiber keys to aid in the identification of individual fibers can be found in the *Textbook of Wood Technology*, 4th Edition, by A. J. Panshin and C. H. deZeeuw, (New York: McGraw-Hill, 1980).

March 1980 Wilfred A. Côté, Editor

CONTENTS

INDEX TO SCIENTIFIC NAMES xi
INDEX TO COMMON NAMES xii
GLOSSARY xv
INTRODUCTION: "FROM WOOD TO PAPER" xix

Natural Fibers of Plant Origin .... plate numbers 1-75

Softwoods 1-24

Sugar Pine	*Pinus lambertiana* Dougl.	1
Western White Pine, Idaho White Pine	*Pinus monticola* Dougl.	2
Eastern White Pine	*Pinus strobus* L.	2
Ray Cell Types in Hard Pines		3
Scotch Pine, Scots Pine	*Pinus sylvestris* L.	4
Red Pine, Norway Pine	*Pinus resinosa* Ait.	4
Monterey Pine	*Pinus radiata* D. Don, *Pinus insignis* Dougl.	5
Shortleaf Pine	*Pinus echinata* Mill.	6
Loblolly Pine	*Pinus taeda* L.	6
Longleaf Pine	*Pinus palustris* Mill.	7
Slash Pine	*Pinus elliottii* Engelm.	7
Ponderosa Pine, Western Yellow Pine	*Pinus ponderosa* Laws.	8
Jack Pine	*Pinus banksiana* Lamb.	8
Pitch Pine	*Pinus rigida* Mill.	9
Western Larch	*Larix occidentalis* Nutt.	9
European Larch	*Larix decidua* Mill.	10
Eastern Larch, Tamarack	*Larix laricina* (DuRoi) K. Koch	10
Red Spruce	*Picea rubens* Sarg.	11
White Spruce	*Picea glauca* (Moench) Voss.	12
Black Spruce	*Picea mariana* (Mill.) B.S.P.	12
Sitka Spruce	*Picea sitchensis* (Bong.) Carr.	13
Engelmann Spruce	*Picea engelmannii* Parry	14
Norway Spruce	*Picea abies* (L.) Karst., *Picea excelsa* Link	15

Douglas-fir	*Pseudotsuga menziesii* (Mirb.) Franco	16
Eastern Hemlock	*Tsuga canadensis* (L.) Carr.	17
Western Hemlock	*Tsuga heterophylla* (Raf.) Sarg.	17
Silver Fir	*Abies alba* Mill.	18
Balsam Fir, Eastern Fir	*Abies balsamea* (L.) Mill.	18
White Fir	*Abies concolor* (Gord. and Glend.) Lindl.	19
Redwood	*Sequoia sempervirens* (D. Don) Endl.	20
Baldcypress	*Taxodium distichum* (L.) Rich.	21
Incense-cedar	*Libocedrus decurrens* Torr.	21
Western redcedar	*Thuja plicata* Donn	22
Port-Orford-Cedar	*Chamaecyparis lawsoniana* (A. Murr.) Parl.	23
Alaska-Cedar	*Chamaecyparis nootkatensis* (D. Don) Spach	23
Parana-Pine	*Araucaria angustifolia* (Bert.) O. Ktze.	24
Hardwoods		25-64
Black Willow	*Salix nigra* Marsh.	25
Bigtooth Aspen	*Populus grandidentata* Michx.	26
European Poplar	*Populus tremula* L.	27
Balsam Poplar	*Populus balsamifera* L.	28
Eastern Cottonwood	*Populus deltoides* Bartr.	29
Black Cottonwood	*Populus trichocarpa* Torr. and Gray	30
Black Walnut	*Juglans nigra* L.	31
Butternut	*Juglans cinerea* L.	32
Shagbark Hickory	*Carya ovata* (Mill.) K. Koch	33
Yellow Birch	*Betula alleghaniensis* Britton	34
Paper Birch, White Birch	*Betula papyrifera* Marsh.	35
European Birch	*Betula verrucosa* Ehrh.	36
Red Alder	*Alnus rubra* Bong.	37
Beech	*Fagus grandifolia* Ehrh.	38
Chestnut	*Castanea dentata* (Marsh.) Borkh.	39
White Oak	*Quercus alba* L.	40
Northern Red Oak	*Quercus rubra* L.	41

American Elm	*Ulmus americana* L.	42
Cucumbertree	*Magnolia acuminata* L.	43
Evergreen Magnolia, Southern Magnolia	*Magnolia grandiflora* L.	44
Yellow-Poplar, Tulip-Poplar	*Liriodendron tulipifera* L.	45
Redbay	*Persea borbonia* (L.) Spreng.	46
Sassafras	*Sassafras albidum* (Nutt.) Nees	47
Sweetgum, Redgum	*Liquidambar styraciflua* L.	48
Sycamore	*Platanus occidentalis* L.	49
Black Cherry	*Prunus serotina* Ehrh.	50
Ailanthus, Tree-Of-Heaven	*Ailanthus altissima* (Mill.) Swingle ...	51
Holly	*Ilex opaca* Ait.	52
Sugar Maple	*Acer saccharum* Marsh.	53
Red Maple	*Acer rubrum* L.	54
Silver Maple	*Acer saccharinum* L.	55
Yellow Buckeye	*Aesculus octandra* Marsh.	56
Basswood	*Tilia americana* L.	57
Loblolly-Bay	*Gordonia lasianthus* (L.) Ellis	58
Kamarere	*Eucalyptus deglupta* Blume	59
Blue Gum	*Eucalyptus globulus* Labill.	60
Water Tupelo, Tupelo-Gum	*Nyssa aquatica* L.	61
Black Tupelo, Blackgum	*Nyssa sylvatica* Marsh.	62
White Ash	*Fraxinus americana* L.	63
Northern Catalpa	*Catalpa speciosa* Warder	64

Miscellaneous Plant Fibers 65-75

Corn	*Zea mays* L.	65
Sugar Cane	*Saccharum officinarum* L.	66
Bamboo	*Cephalostachyum pergracile* Munro	67
Wheat	*Triticum* sp.	68
Rice	*Oryza sativa* L.	69
Esparto Grass	*Stipa tenacissima* L.	70
Manila Hemp	*Musa textilis* Née	71
Jute	*Corchorus* sp.	71
Hemp	*Cannabis sativa* L.	72

Sisal, Henequen	*Agave* sp.	72
Paper-Mulberry	*Broussonetia papyrifera* Vent.	73
Pineapple	*Ananas* sp.	73
Ramie, China Grass	*Boehmeria nivea* (L.) Gaud.	74
Flax	*Linum usitatissimum* L.	74
Cotton	*Gossypium* sp.	75
Kapok	*Ceiba pentandra* Gaertn.	75

Natural Fibers of Animal Origin

Wool 76
Silk 76

Mineral Fiber

Asbestos 77

Man-Made Fibers

Rayon 78
Nylon 78
Cross-Sections 79
- Regular rayon
- High tensile rayon ("Avril")
- "Dacron"
- XL HiStrength rayon
- Secondary acetate
- "Orlon"
- Nylon

INDEX TO SCIENTIFIC NAMES

Abies alba 18
balsamea 18
concolor 19
Acer rubrum 54
saccharinum 55
saccharum 53
Aesculus octandra 56
Agave sp. 72
Ailanthus altissima 51
Alnus rubra 37
Ananas sp. 73
Araucaria angustifolia 24
Betula alleghaniensis 34
papyrifera 35
verrucosa 36
Boehmeria nivea 74
Broussonetia papyrifera 73
Cannabis sativa 72
Carya ovata 33
Castanea dentata 39
Catalpa speciosa 64
Ceiba pentandra 75
Cephalostachyum pergracile . 67
Chamaecyparis lawsoniana ... 23
nootkatensis 23
Corchorus sp. 71
Eucalyptus deglupta 59
globulus 60
Fagus grandifolia 38
Fraxinus americana 63
Gordonia lasianthus 58
Gossypium sp. 75
Ilex opaca 52
Juglans cinerea 32
nigra 31
Larix decidua 10
laricina 10
occidentalis 9
Libocedrus decurrens 21
Linum usitatissimum 74
Liquidambar styraciflua 48
Liriodendron tulipifera 45
Magnolia acuminata 43
grandiflora 44
Musa textilis 71
Nyssa aquatica 61
sylvatica 62
Oryza sativa 69
Persea borbonia 46
Picea abies 15
engelmannii 14
excelsa (*see* P. abies) 15
glauca 12
mariana 12
rubens 11
sitchensis 13
Pinus banksiana 8
elliottii 7
Pinus echinata 6
insignis (*see* P. radiata) . 5
lambertiana 1
monticola 2
palustris 7
ponderosa 8
radiata 5
resinosa 4
rigida 9
strobus 2
sylvestris 4
taeda 6
Platanus occidentalis 49
Populus balsamifera 28
deltoides 29
grandidentata 26
tremula 27
trichocarpa 30
Pseudotsuga menziesii 16
Prunus serotina 50
Quercus alba 40
rubra 41

Saccharum officinarum 66
Salix nigra 27
Sassafras albidum 47
Sequoia sempervirens 20
Stipa tenacissima 70
Taxodium distichum 21
Thuja plicata 22
Tilia americana 57
glabra (*see* T. americana) 57
Triticum sp. 68
Tsuga canadensis 17
heterophylla 17
Ulmus americana 42
Zea mays 65

INDEX TO COMMON NAMES

Ailanthus 51
Alaska-cedar 23
Alder, red 37
American elm 42
Asbestos 77
Ash, white 63
Aspen, bigtooth 26
Baldcypress 21
Balsam fir 18
Balsam poplar 28
Bamboo 67
Basswood 57
Bay (*see* redbay) 46
Beech 38
Bigtooth aspen 26
Birch, European paper (white) 35
yellow 34
Black cherry 50
Black cottonwood 30
Blackgum 62
Black spruce 12
Black tupelo (*see* blackgum) . 62
Black walnut 31
Black willow 25
Blue gum 60
Buckeye (*see* yellow buckeye) 56
Butternut 32
Catalpa, northern 62
Cedar
Alaska-cedar 23
Incense-cedar 21
Port-Orford-cedar 23
Western redcedar 22
Cherry (*see* black cherry) 50
Chestnut 39
China grass (*see* ramie) 74
Corn 65
Cotton 75
Cottonwood, black 30
eastern 29
Cucumbertree 43
Cypress (*see* baldcypress) 21
Douglas-fir 16
Eastern cottonwood 29
Eastern fir (*see* balsam fir) 18
Eastern hemlock 17
Eastern larch 10
Eastern white pine 2
Elm, American 42
Engelmann spruce 14
Esparto grass 70
European birch 36
European larch 10
European poplar 27

Evergreen magnolia 44
Fir, balsam 18
silver 18
white 19
(*see* also Douglas-fir) 16
Flax 74
Grass, esparto 70
Gum (*see* blackgum) 62
(*see* redgum) 48
(*see* water tupelo) 61
Hemlock, eastern 17
western 17
Hemp, "true" 72
Manila 71
Henequen 72
Hickory, shagbark 33
Holly 52
Idaho white pine 2
Incense-cedar 21
Jack pine 8
Jute 71
Kamarere 59
Larch, eastern 10
European 10
western 9
Linen (*see* flax) 74
Loblolly pine 6
Longleaf pine 7
Magnolia
(*see* evergreen magnolia) .. 44
(*see* cucumbertree) 43
Manila hemp 71
Maple (*see* red maple) 54
(*see* silver maple) 55
(*see* sugar maple) 53
Monterey pine 5
Northern catalpa 64
Northern red oak 41
Norway pine 4
Norway spruce 15
Nylon 78, 79
Oak, northern red 41
white 40
Paper birch 35
Paper-mulberry 73
Parana-pine 24
Pineapple 73
Pine, eastern white 2
Idaho white 2
jack 8
loblolly 6
longleaf 7
Monterey 5
Norway 4
pitch 9
ponderosa 8
red 4
Scotch 4
Scots 4
shortleaf 6
slash 7
sugar 1
western white 2
(*see* also Parana-pine) ... 24
Pitch pine 9
Ponderosa pine 8
Poplar, balsam 28
(*see* European) 27
(*see* Aspen, bigtooth) 26
(*see* Cottonwood, black) 30
(*see* Cottonwood, eastern) 29
(*see* also yellow-poplar) .. 45
Port-Orford-cedar 23
Ramie 74
Rayon 78, 79
Red alder 37
Redbay 46
Redgum 48
Red maple 54
Red oak
(*see* northern red oak) 41
Red pine 4
Red spruce 11

Redwood 20
Rice 69
Sassafras 47
Scotch pine 4
Scots pine 4
Shagbark hickory 33
Shortleaf pine 6
Silk 76
Silver fir 18
Silver maple 55
Sisal 72
Sitka spruce 13
Slash pine 7
Southern magnolia
(*see* evergreen magnolia) 44
Southern yellow pines
(*see* loblolly pine) 6
(*see* longleaf pine) 7
(*see* shortleaf pine) 6
(*see* slash pine) 7
Spruce, black 12
Engelmann 14
Norway 15
Spruce, red 11
Sitka 13
white 12
Straw (*see* rice) 69
(*see* wheat) 68
Sugar cane 66
Sugar maple 53
Sugar pine............................. 1
Sweetgum (*see* redgum) 48
Swiss-pine (*see* silver fir) 18
Sycamore 49
Tamarack (*see* eastern larch) . 10
Tree-of-Heaven
(*see* ailanthus) 51
Tulip-poplar
(*see* yellow-poplar) 45
Tupelo (*see* water tupelo) 61
Tupelo-gum (*see* water tupelo) 61
Walnut, black 31
white (*see* butternut) 32
Water tupelo 61
Western hemlock 17
Western larch 9
Western redcedar 22
Western white pine 2
Western yellow pine 8
Wheat 68
White ash 63
White fir 19
White oak 40
White spruce 12
Willow, black 25
Wool 76
Yellow birch 34
Yellow buckeye 56
Yellow-poplar 45

GLOSSARY

ANNULAR VESSEL: a vessel with annular thickenings on the inner surface of its otherwise relatively thin wall; the rings may be close-spaced or at wide and frequently at irregular intervals.

BAST FIBER: a fiber from the phloem portion of the vascular tissue of a plant, as distinguished from its xylary portion.

BORDERED PIT: a pit with an overhanging margin, *i.e.*, a pit in which the cavity becomes abruptly constricted during the thickening of the secondary wall.

ELEMENT: a specific cell type.

EPIDERMAL CELL: a cell from the epidermis of a plant; the epidermis is the tegumentary (covering) tissue of the plant.

FIBER: a general term for a narrow elongated cell with tapering ends; different types are recognized according to wall thickness, width of lumen, and nature of the pitting.

GROUND TISSUE: the thin-walled parenchyma between the vascular bundles in a monocotyledonous stem.

HELICAL THICKENING: spiralled ridgelike thickening of the inner wall of a cell.

INTERTRACHEID PITS: pits of pit pairs in the wall of a tracheid leading to a contiguous tracheid before the wood was pulped.

INTERVESSEL PITS: pits of pit pairs in the wall of a vessel leading to a contiguous vessel before the wood was pulped.

LONGITUDINAL: directed or extending lengthwise, as along the long axis of a plant stem or parallel to the grain of wood.

LONGITUDINAL PARENCHYMA: strands of parenchymatous cells extending along the grain in wood.

LUMEN: the cavity of a cell enclosed by its wall; viewed laterally, it is canal-shaped when the cell is elongated.

PARENCHYMA: collective noun designating relatively short cells with simple pits; they are usually quite thin-walled and often remain alive for a considerable time.

PERFORATION: the opening (or openings) between two vessel segments.

PIT: a recess in the secondary wall of a cell spanned at its outer end by a closing membrane (pit membrane); pits result through failure of the secondary wall to form at these points.

RAY: a ribbon-like band of cells extending radially in the vascular tissue of a tree so oriented that the face of the ribbon is exposed in a radial section.

RAY-CELL CONTACT AREA: the pitted area on the wall of a longitudinal element where a ray cell was originally in contact with it.

RAY-CONTACT AREA: the pitted area on the wall of a longitudinal element where a ray was originally in contact with it.

RAY PARENCHYMA: collective term for ray parenchymatous cells with simple pits, and with living contents as long as they are in the sapwood of a tree.

RAY TRACHEID: a ray cell with bordered pits and devoid of living contents at maturation (*see* ray parenchyma).

RETICULATE: netlike; said of a cell with either openings or variations in wall thickness which have a netlike pattern.

SCALARIFORM: ladder-like; said of linear pits arranged transversely in parallel, and of perforations with transverse parallel bars.

SCALARIFORM PERFORATION: a perforation spanned by two to many transverse bars in a ladder-like arrangement.

SCALARIFORM PITTING: said of linear pits arranged scalariformly (*see* scalariform).

SIMPLE PERFORATION: a round or oval perforation without cross bars.

SPRINGWOOD: the wood formed during the early part of a growing season; the inner, usually less dense portion of an annual increment; also called earlywood.

SPRINGWOOD TRACHEID: a tracheid in the springwood of softwoods; it has a larger radial diameter, a thinner wall, and wider lumen than a summerwood tracheid (*see* summerwood tracheid).

SPRINGWOOD VESSEL SEGMENT: a vessel segment in the springwood of a ring porous hardwood; it is appreciably wider than a summerwood vessel segment (*see* summerwood vessel segment).

STOMATE: the opening and the adjacent specialized guard cells in the epidermis of a plant (*see* epidermal cell).

SUMMERWOOD: the wood formed during the latter part of a growing season; the outer, usually denser portion of an annual increment; also called latewood.

SUMMERWOOD TRACHEID: a tracheid in the summerwood of softwoods; it has a smaller radial diameter, thicker wall, and narrower lumen than a springwood tracheid (*see* springwood tracheid).

SUMMERWOOD VESSEL SEGMENT: a vessel segment in the summerwood of a ring porous hardwood; it is appreciably narrower than a springwood vessel segment (see springwood vessel segment).

TRACHEID: a narrow elongated cell with tapering ends and numerous bordered pits.

TYLOSIS (plural—TYLOSES): a balloon-like or saclike structure sometimes present as an inclusion in a vessel segment.

VASCULAR TRACHEID: specialized cells in certain hardwoods, similar in size, shape and arrangement to the small vessel segments but differing from them in being imperforate at the ends.

VASICENTRIC TRACHEID: a relatively short, crooked fibrous cell with many bordered pits.

VESSEL ELEMENT: one of the perforated units of a vessel; a vessel consists of a longitudinal series of such units; also called vessel member and vessel segment.

INTRODUCTION

From Wood to Paper

Papermaking as an art is now nearly twenty centuries old. From its beginnings in China, it took many centuries to reach Western Europe where paper was made about A.D. 1100. Throughout this period the raw material used was the inner bark of the Mitsumata tree (paper mulberry), old rags, hemp, and other vegetable fiber. Even as the art evolved into technology, the same sources of fiber were used. They were all cellulosic, but did not include wood.

The first wood pulp was produced in Germany in the 1840's. It was mechanical pulp or groundwood. Some years later chemical pulping was developed so that approximately one hundred years ago wood emerged as a major source of fiber for papermaking. In spite of its late arrival, the past century has seen the use of wood grow enormously until now more than 90% of all paper is based on wood fiber.

This photomicrographic atlas of papermaking fibers reflects the historical aspects as well as the modern reality of the industry. Seventy-five wood species are represented. Yet, there remain the important agricultural waste fibers from sugar cane, wheat and rice straw; and from the widely used bamboo, hemp, esparto grass, flax, and cotton. Animal fibers are also included: wool and silk. Asbestos represents the category of mineral fiber. Finally, man-made or chemical fiber is also illustrated.

The xylem or wood portion of both gymnosperms (conifers) and arborescent angiosperms is the source of wood fiber for papermaking. In both, typical softwoods and hardwoods, cellulose makes up $42 \pm 2\%$ of the xylem. The conifers contain 30% hemicelluloses as do the angiosperms. The lignin content of softwoods averages 27 to 34% while in the hardwoods it ranges from 23 to 29%.

Because of the importance of wood as a raw material, this short introduction has been included in this publication to illustrate how wood is converted into paper, structurally or morphologically. For this purpose, chemical "pulping" of small wood pieces was carried out after scanning electron micrographs were recorded to show the three-dimensional structure of wood blocks (Figures 1, 2, 8 and 9). Then additional micrographs were made to illustrate how a pulpwood chip separates into cellulosic fibers (Figures 3, 4 and 10), with enlarged views of such fibers before they are re-assembled into paper (Figures 5, 11-15). Finally, photographs of the paper sheet were made (Figures 6, 7, 16 and 17).

Since there are two broad categories of trees utilized as sources of wood fiber, both need to be included. The choice of a representative softwood is not very critical since the structure of one is quite similar to another. Tracheids are the predominant cell type, making up more than 90% of the volume of a piece of coniferous wood. These are very long, slender cells ranging from approximately 2.5 to 5 mm. in length. Their width is only 20 to 50 micrometers. These long cells are oriented longitudinally in the tree and appear in the wood as shown in Figure 1.

The other major cell type in the wood of conifers is ray parenchyma. These cells make up the wood rays (Figure 1) which can be seen as ribbons of tissue radiating from the center of the tree between radial files of tracheids. In some species there are so-called ray tracheids which are also conductive cells like the longitudinal tracheids. They are found with the ray parenchyma in the rays. Both are very short cells compared with the longitudinal elements.

The structure of hardwoods is more complex and variable than that of softwoods. The selection of a typical hardwood is not easily made. There are more cell types and their distribution is not as uniform as in the conifers. Each species exhibits its own characteristic pattern or design so if oak is selected to represent the angiosperms, it would not resemble poplar, birch, or redgum.

The presence of vessels characterizes hardwoods (Figures 8 and 9). Individual vessel elements may be of very large diameter, but their lengths are much less than for coniferous tracheids. Vessel elements are open-ended to connect with others and thus provide a long conductive path vertically in the tree. In some species vessels are very large in the springwood and then may gradually be reduced in size as the growing season progresses. In others there is an abrupt reduction in vessel size after the first few vessels are produced at the beginning of wood formation in the spring. In yet others, vessels are of more uniform but smaller dimensions.

Other cell types found in hardwood are libriform fibers which are long, slender, small-lumened elements; vasicentric tracheids and fiber-tracheids, which appear to be transition elements since they are larger than fibers but smaller than vessels. The proportion of each type and their distribution within the structure of wood is highly variable. As can be seen in Figure 8, not only are rays present in hardwoods, but they may be more prominent features than in softwoods. Rays can be one cell or many cells wide and occasionally very high. As in softwoods, ray parenchyma cells (Figure 10) are short, mostly rectangular in shape, and much smaller than the longitudinal elements.

Because of the great variation in cell size throughout the hardwood category, only a wide range of cell dimensions could be suggested. It would be preferable to refer to individual species in this manual and to note the magnification listed for each element. In addition, a micrometer scale has been placed on many of the scanning electron micrographs to provide a direct visual comparison of size.

Before wood can be separated into individual cells, the middle lamella region (Figures 2 and 9) between elements must be attacked chemically (or mechanically as in groundwood). This intercellular zone contains a high concentration of lignin. Pulping, or maceration in small laboratory quantities, provides for the removal of lignin from this region and from the cell wall itself, thus producing separated cellulosic fibers after washing.

In Figures 3 and 4 for softwoods and Figure 10 for hardwoods, the effect of pulping can be seen as the cells are essentially freed from each other but still remain in approximately the position they held in the solid wood. They are then separated as shown in Figures 5 and 11 through 15. Note the structural differences between softwood and hardwood cells. The nature of the pits, gaps in the cell wall which provide for inter-cell communication, and of other sculpturing of the wall is intricate and reveals something about the function of the cell.

The fibers, the generic name given to all wood cells used in papermaking regardless of their anatomical features or origin, are then rearranged randomly and bonded to each other to form a sheet, a product having much different properties from the original wood. The detail of paper structure can be studied in micrographs of paper surface and paper cross sections (Figures 6, 7, 16 and 17). Note that in these uncoated and unfilled paper sheets it is still possible to detect some of the structure of the individual wood fibers.

Papermaking Fibers is designed to help identify the source of fiber used in papermaking. To use these photomicrographs effectively it would be advantageous to try to relate the cell type to the original wood and that cell's function or role. A useful reference on wood structure and identification is *Wood—Structure and Identification*, 2nd Edition, by H. A. Core, W. A. Côté and A. C. Day, Syracuse University Press, 1979. For nonwoody fibers there is much less published information.

Figure 1. Scanning electron micrograph of a cube of eastern white pine microtomed on three surfaces. Note the arrangement of longitudinal tracheids (tr) in radial files and the structure of the rays. In this species resin canals (rc) are rather prominent features.

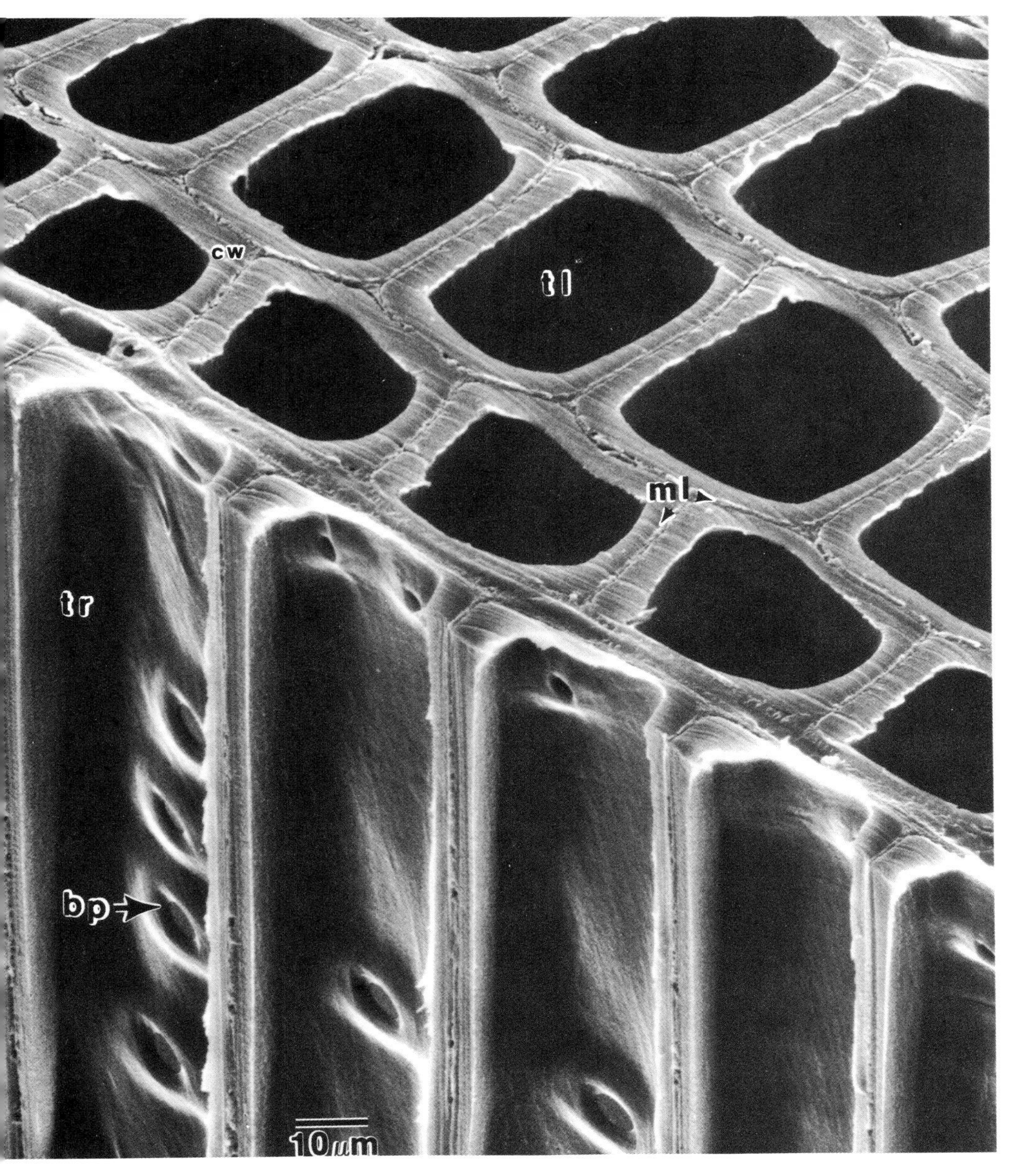

Figure 2. Scanning electron micrograph of a portion of a cube of eastern white pine such as in Figure 1. At this higher magnification, the cell walls (cw) and the middle lamella (ml) can be seen. Removal of the middle lamella lignin by pulping frees the cells from each other as can be seen in Figures 3 and 4. The tracheid lumens (tl) may be viewed in cross section or in a longitudinal exposure indicated here as tr. It is also possible to see the bordered pits (bp) on the walls of the tracheids. These gaps in the secondary wall provide communication between tracheids which have closed ends as can be seen in Figure 4.

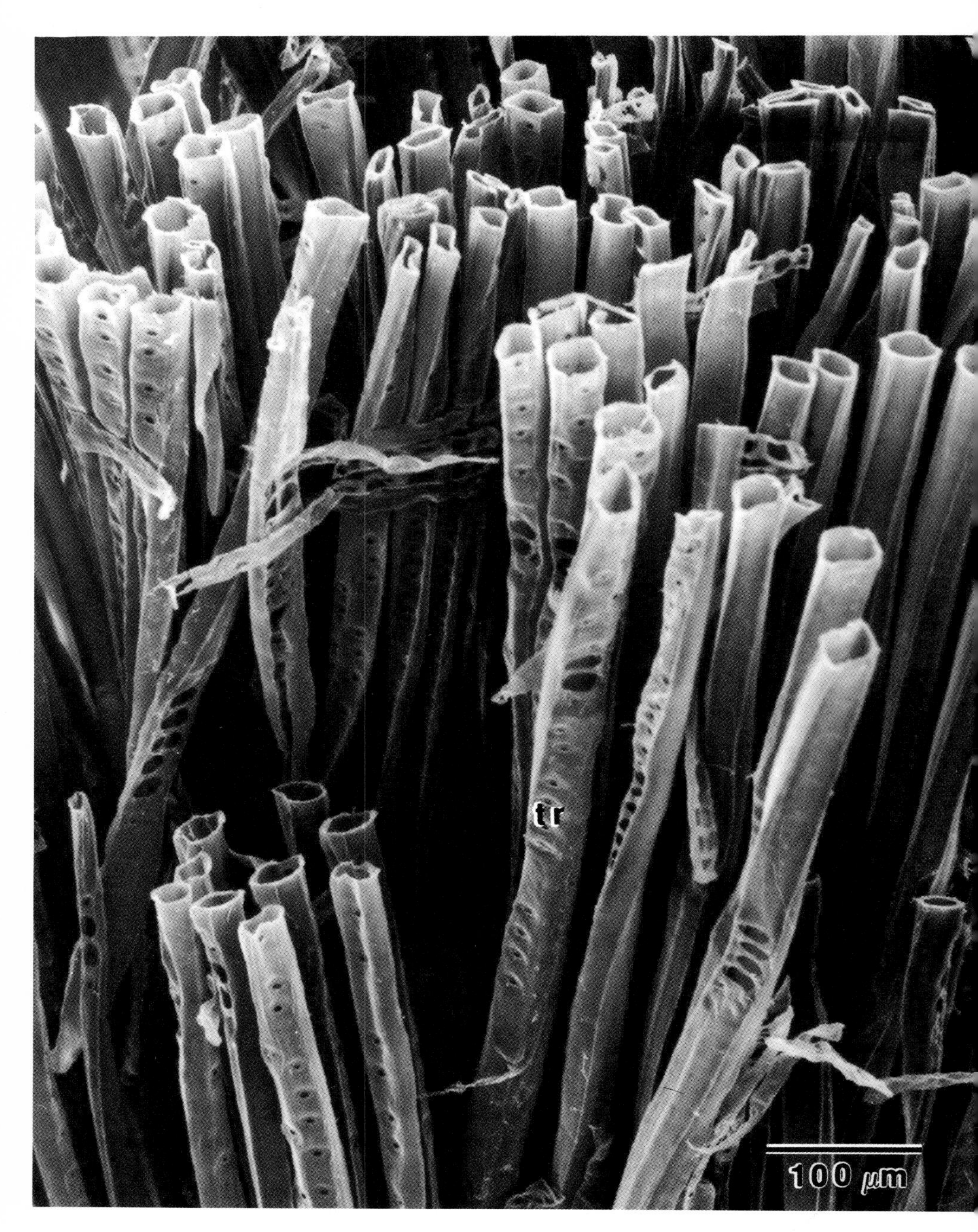

Figure 3. Scanning electron micrograph of a cluster of eastern white pine elements after pulping or maceration of a small cube such as in Figure 1. Note that the tracheid ends have been cut off in microtoming.

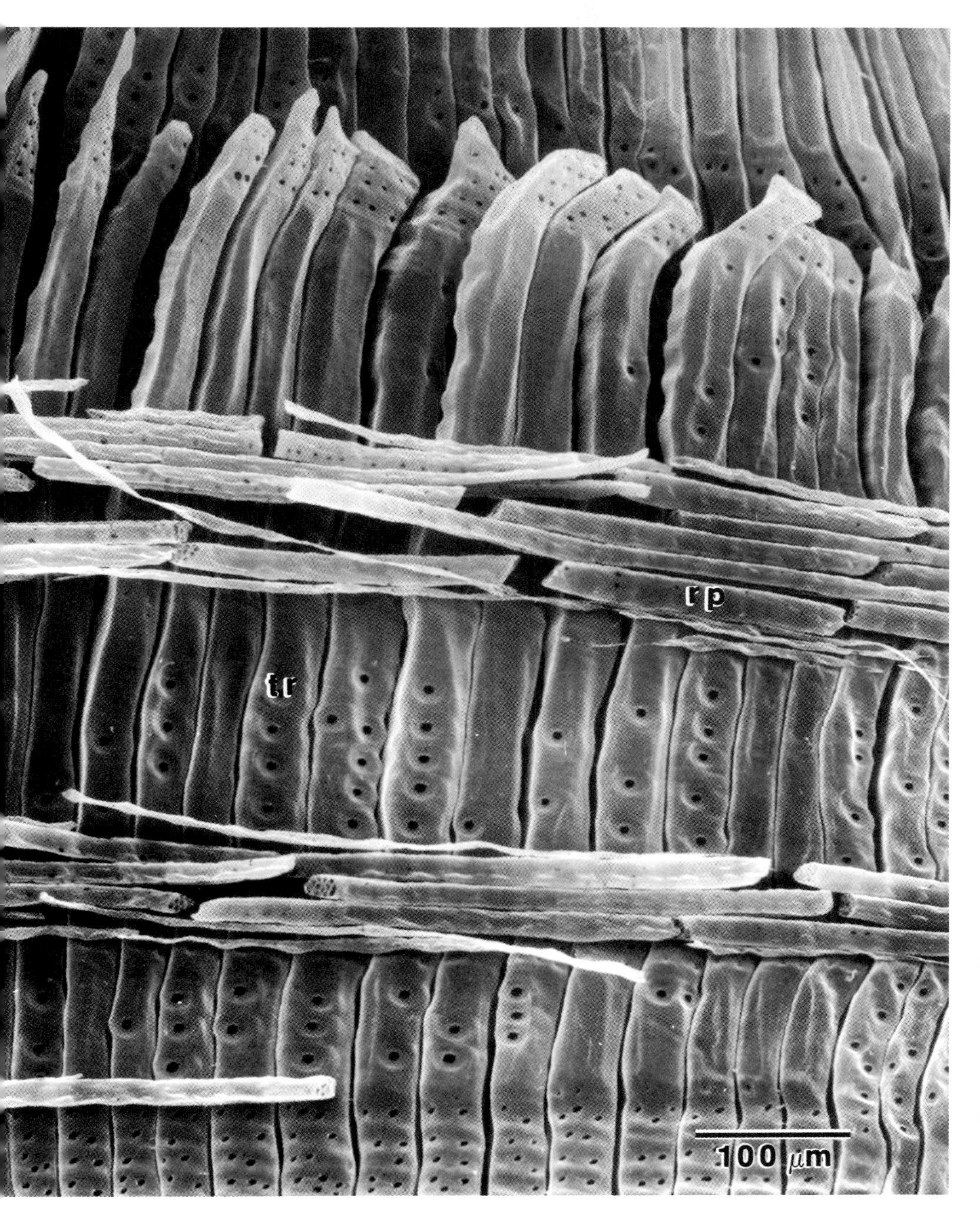

Figure 4. Scanning electron micrograph of pulped or macerated Douglas-fir showing the longitudinal tracheids (tr) with their closed ends and ray parenchyma cells (rp) arranged in ribbon-like aggregations.

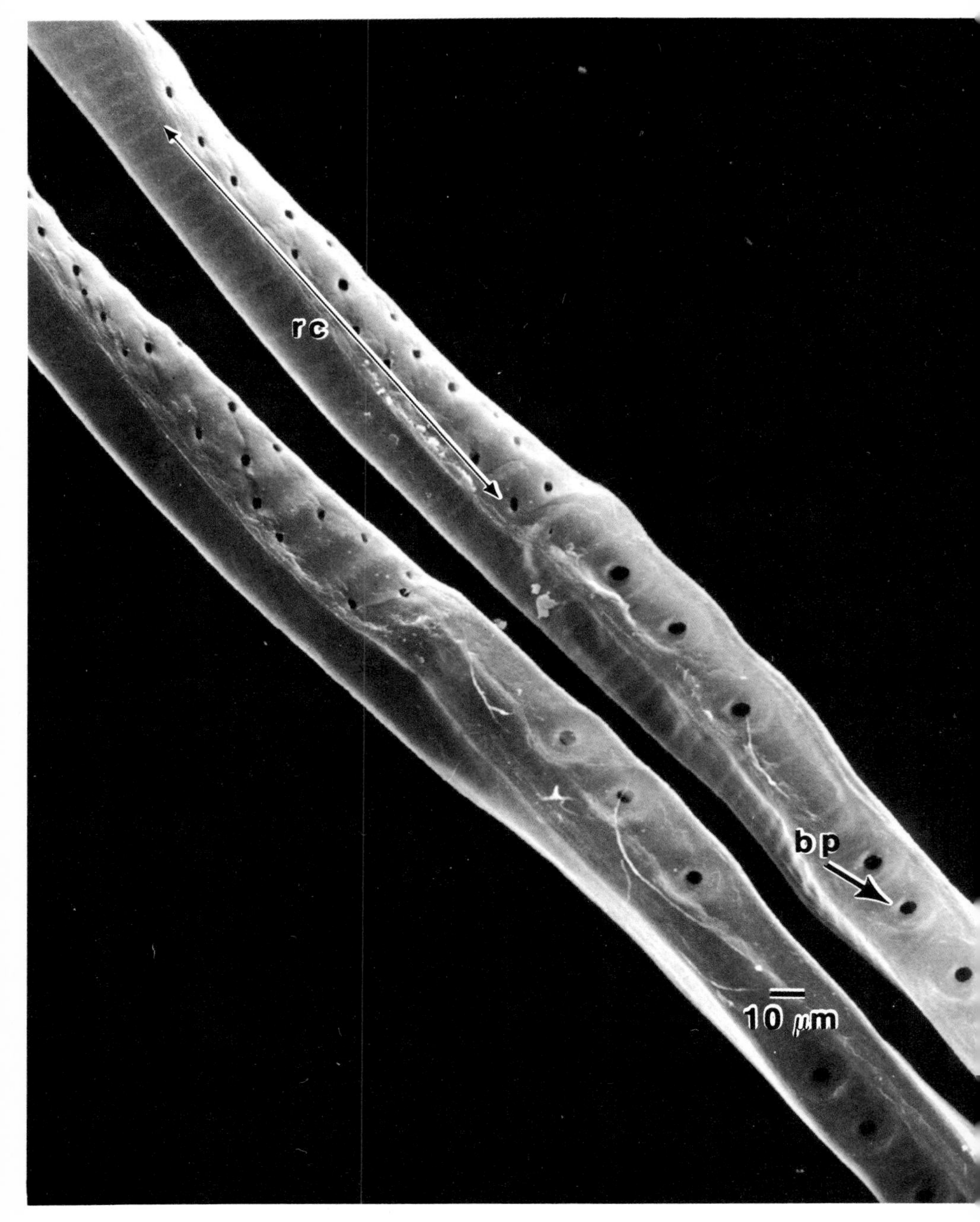

Figure 5. Scanning electron micrograph of portions of two Douglas-fir tracheids showing bordered pits (bp) and ray contact areas (rc). The pits in the ray contact area provide communication between longitudinal tracheids and ray parenchyma cells.

Figure 6. Scanning electron micrograph of paper surface showing random arrangement of coniferous tracheids in the sheet.

Figure 7. Scanning electron micrograph at higher magnification showing both the paper surface and the cross section which was cut with a razor blade. Note the flattened or collapsed nature of the fibers.

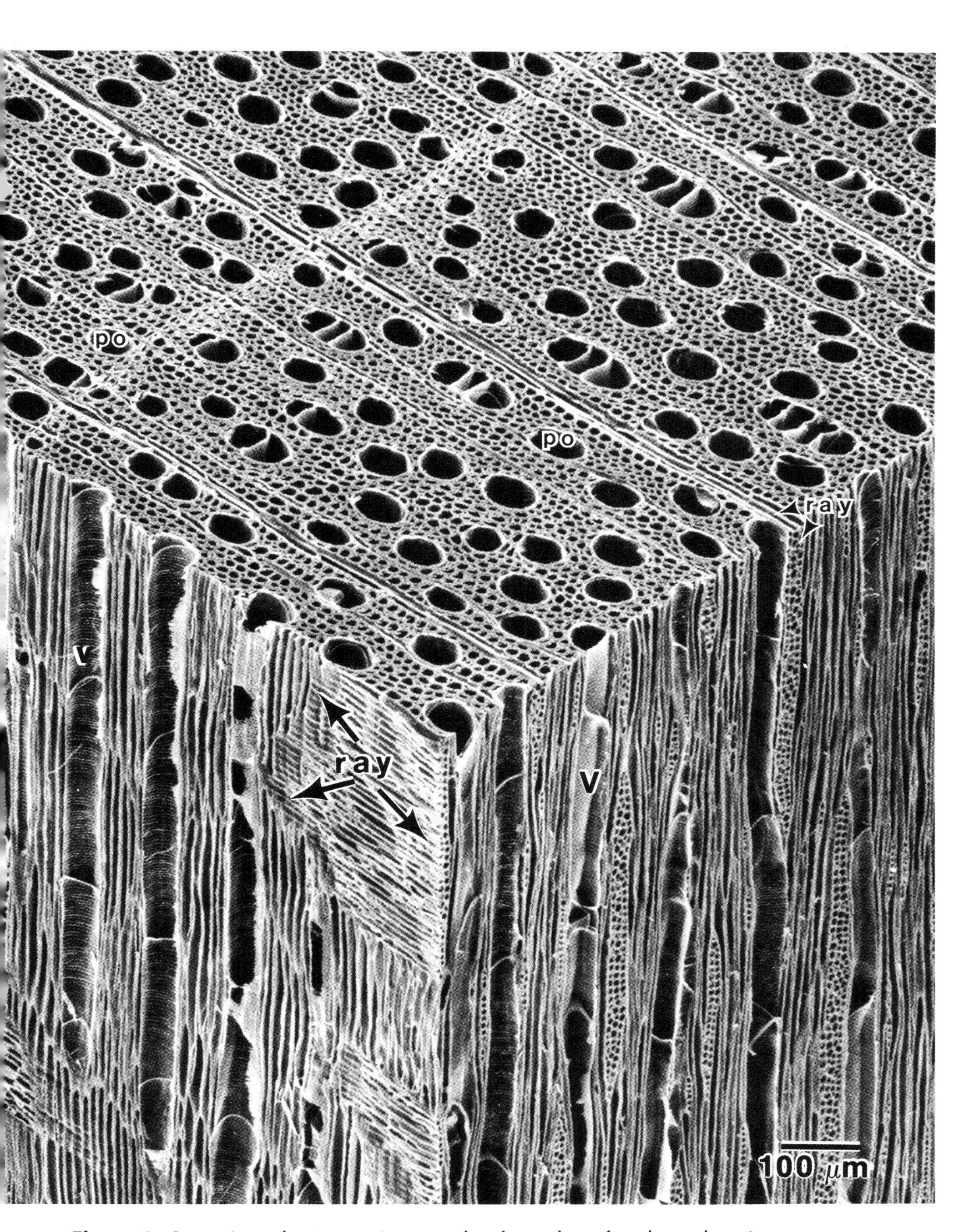

Figure 8. Scanning electron micrograph of a cube of red maple microtomed on three surfaces. Compare the structure of this hardwood with that of a softwood, eastern white pine, in Figure 1. Vessels (v) characterize hardwoods. When viewed on the transverse surface, vessels are called pores (po). The rays in hardwoods, as can be seen from this micrograph, are frequently higher and wider than in softwoods.

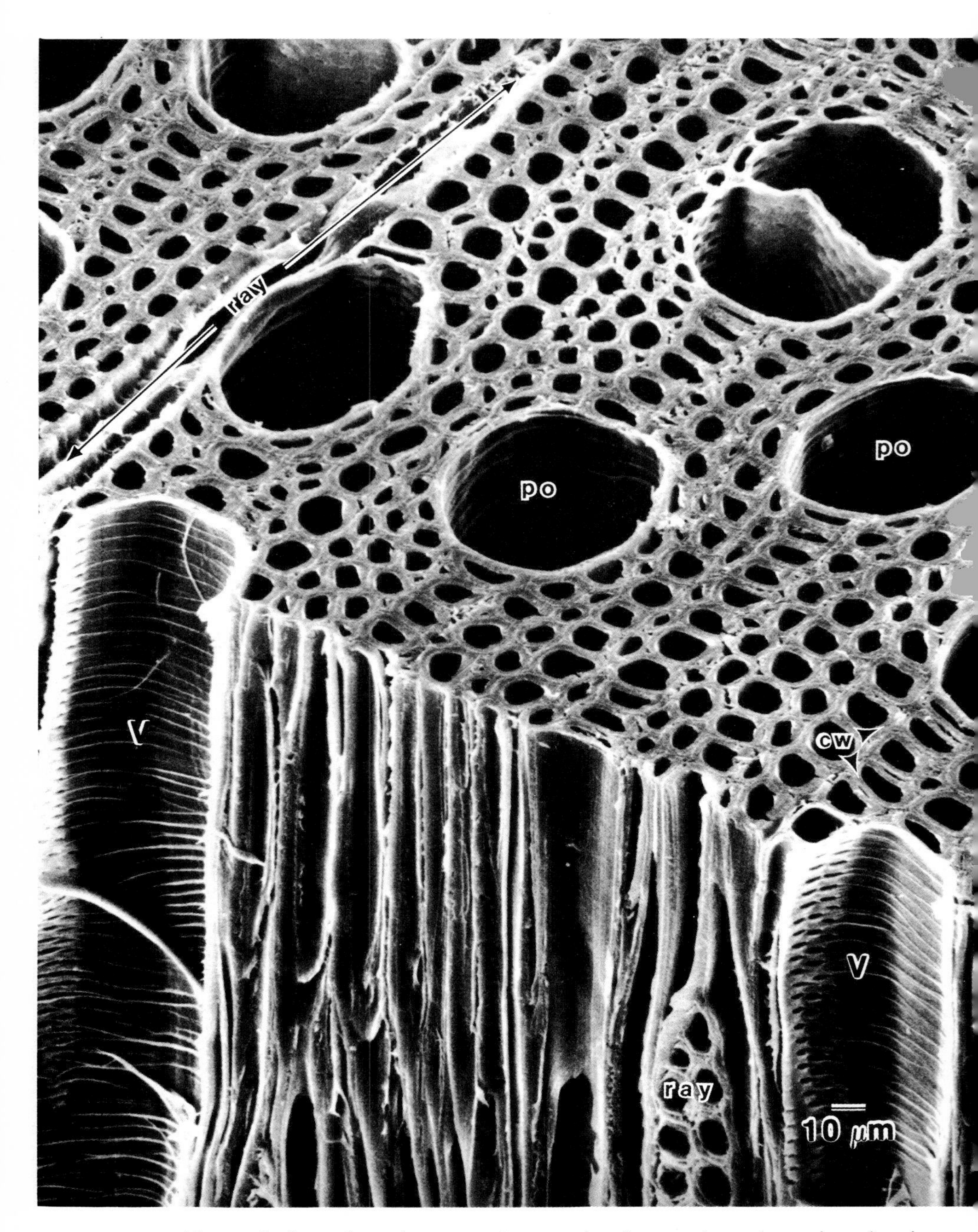

Figure 9. Scanning electron micrograph of a portion of a cube of red maple at a higher magnification than in Figure 8. The walls of the vessels (v) are sculptured with helical thickenings. The cell walls (cw) can be seen more clearly at this magnification, but to make the middle lamella more visible would require still greater enlargement. The ray structure on both the transverse and the tangential faces is clearly resolved.

Figure 10. Scanning electron micrograph of pulped or macerated redgum in which the elements have been carefully left undisturbed following treatment. The individual vessel elements (ve) and the ray parenchyma cells (rp) are more distinguishable after maceration.

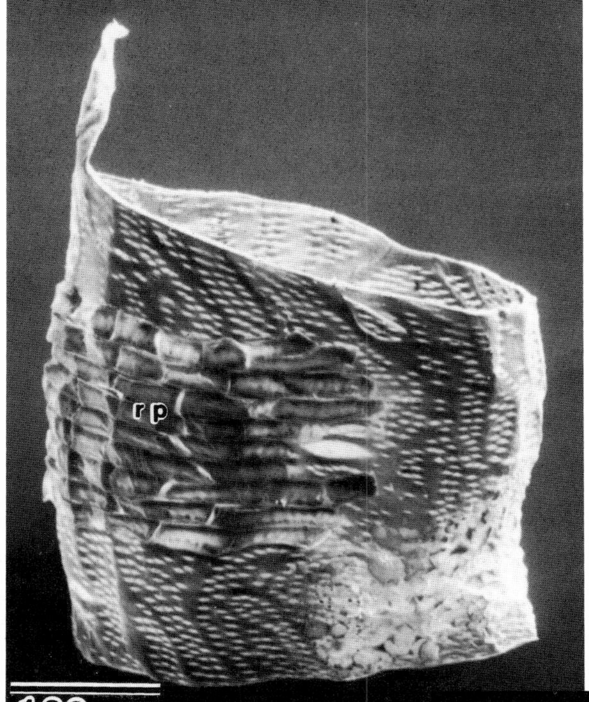

Figure 11. Scanning electron micrograph of red oak vessel element. Patterns of pitting suggest positions of the cells which had been in contact with this element. Note that some ray parenchyma cells (rp) still adhere to the vessel element, blocking a view of the underlying pits.

Figure 12. Scanning electron micrograph of a vessel element (ve) and an attached fiber (f) from eastern cottonwood. There is a prominent ray contact area (rc) on the vessel element.

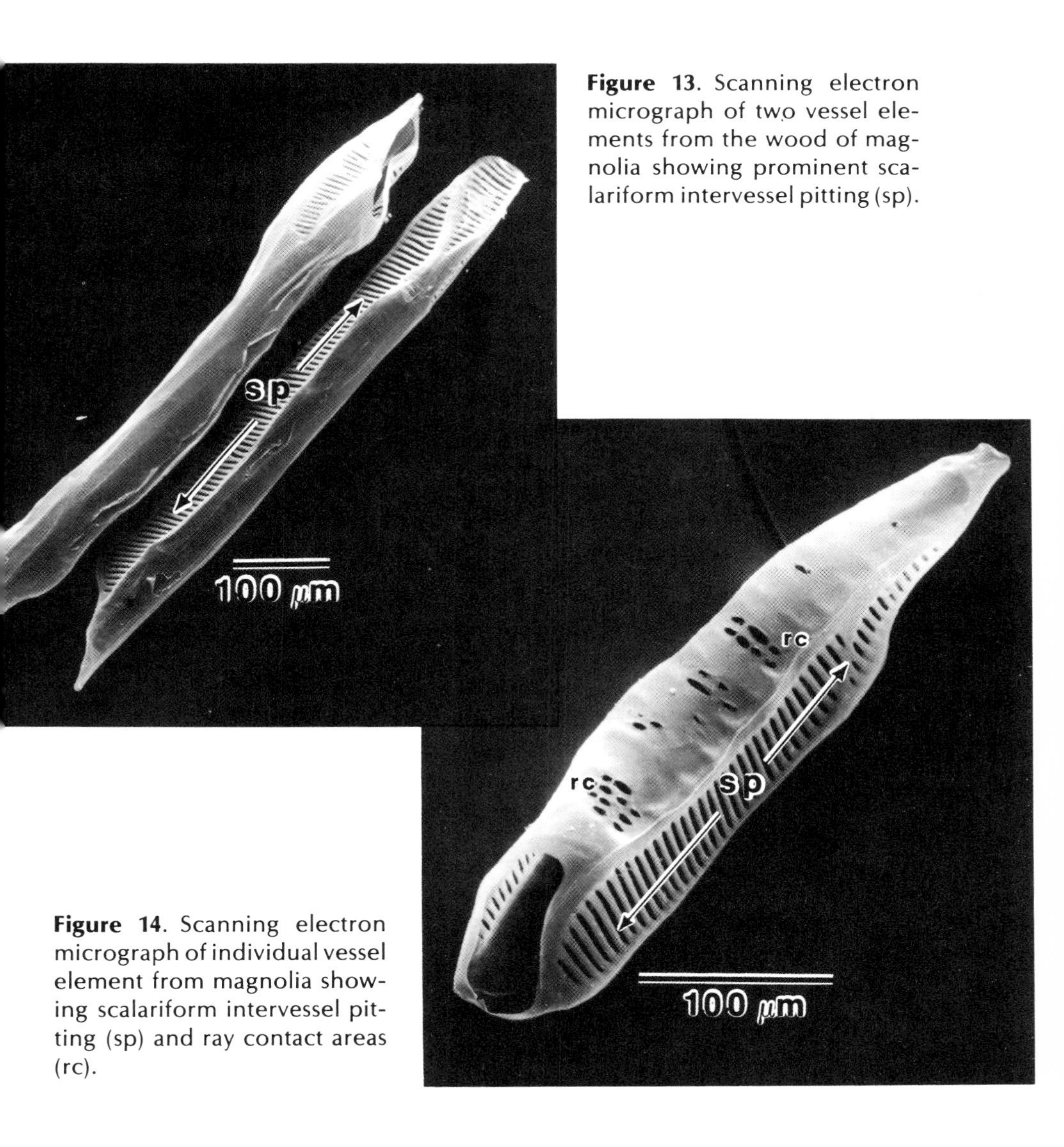

Figure 13. Scanning electron micrograph of two vessel elements from the wood of magnolia showing prominent scalariform intervessel pitting (sp).

Figure 14. Scanning electron micrograph of individual vessel element from magnolia showing scalariform intervessel pitting (sp) and ray contact areas (rc).

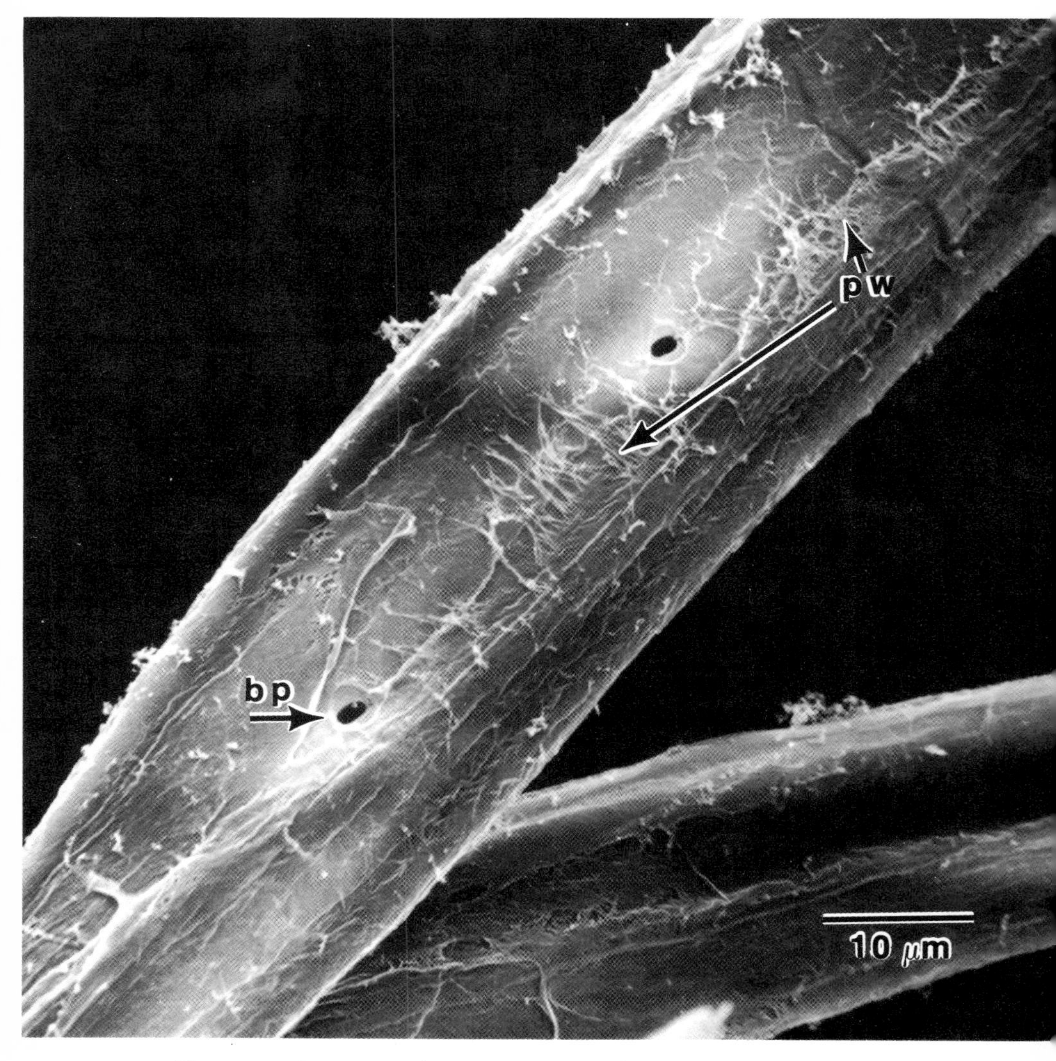

Figure 15. Scanning electron micrograph of the surface of a portion of yellow-poplar fibers revealing the fine network of primary wall (pw) fragments adhering to the secondary wall. Note bordered pits (bp).

Figure 16. Scanning electron micrograph of paper surface showing the arrangement of hardwood cells including vessel elements.

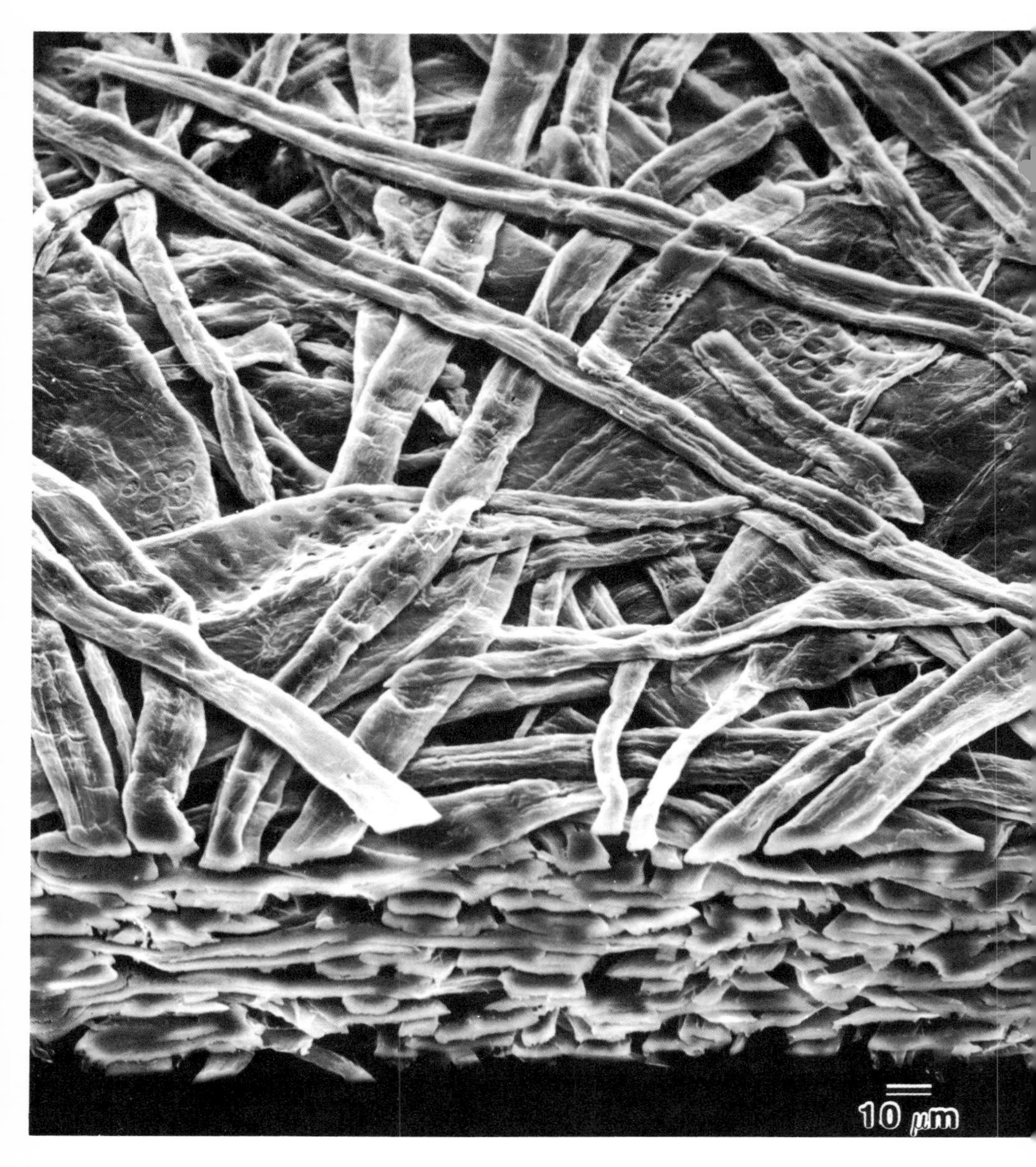

Figure 17. Scanning electron micrograph showing both the paper surface and the cross section which was cut with a razor blade. The flattened nature of the collapsed fibers can be seen on the cross section.

PAPERMAKING FIBERS

SUGAR PINE
Pinus lambertiana Dougl.

1. Springwood longitudinal tracheid, 95X, radial view
   a. Intertracheid pits: bordered pits of pit pairs leading to a contiguous longitudinal tracheid
   b. Ray-contact area

2. Portion of a longitudinal tracheid, 185X
   c. Ray-tracheid pits: small bordered pits of pit pairs leading to a ray tracheid
   d. Ray-parenchyma pits: large, window-like pits of pit pairs leading to ray parenchyma
   e. Intertracheid pits: bordered pits of pit pairs leading to a contiguous longitudinal tracheid

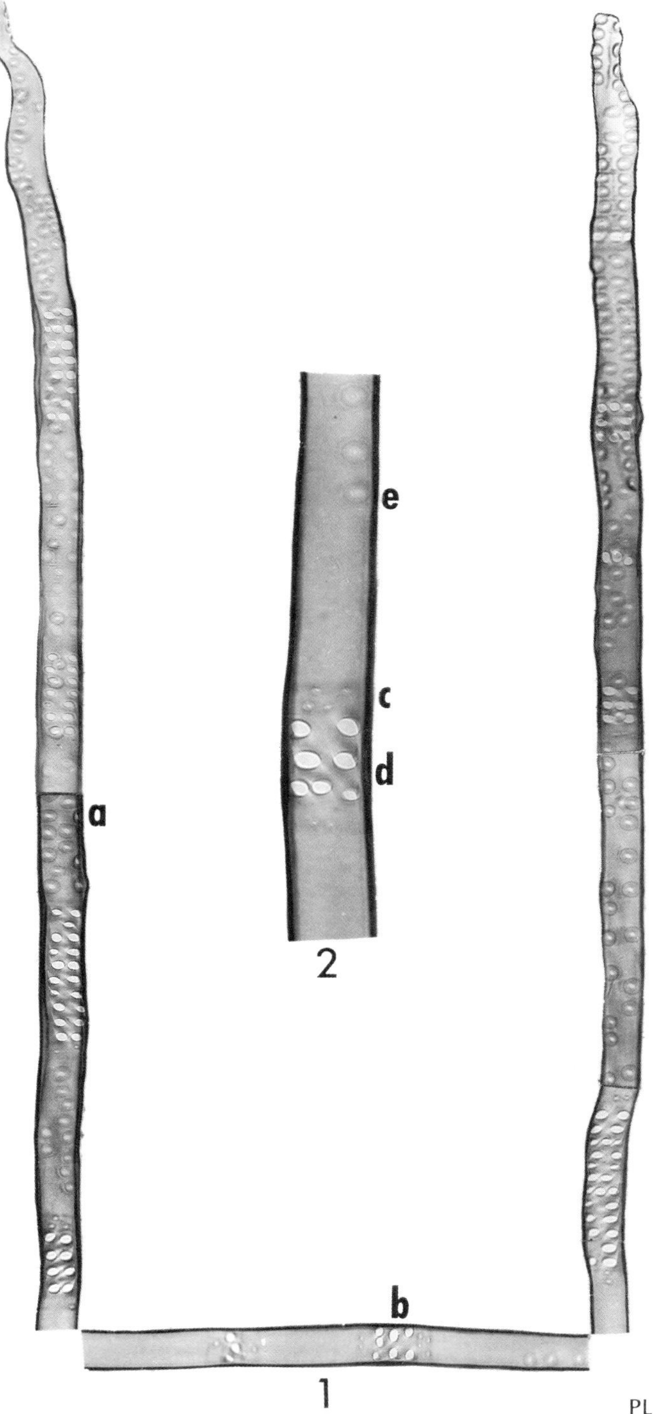
a
b
c
d
e
1
2

WESTERN WHITE PINE, IDAHO WHITE PINE

Pinus monticola Dougl.

1. Springwood longitudinal tracheid, 95X, radial view
   a. Intertracheid pits: bordered pits of pit pairs leading to a contiguous longitudinal tracheid
   b. Ray-contact area

2. Portion of a longitudinal tracheid, 185X
   c. Ray-tracheid pits: small, bordered pits of pit pairs leading to a ray tracheid
   d. Ray-parenchyma pits: large, window-like pits of pit pairs leading to a ray parenchyma

EASTERN WHITE PINE

Pinus strobus L.

1. Springwood longitudinal tracheid, 95X, radial view
   a. Intertracheid pits: bordered pits of pit pairs leading to a contiguous longitudinal tracheid
   b. Ray-contact area

2. Portion of a longitudinal tracheid, 185X
   c. Ray-tracheid pits: small, bordered pits of pit pairs leading to a ray tracheid
   d. Ray-parenchyma pits: large, window-like pits of pit pairs leading to ray parenchyma
   e. Intertracheid pits: bordered pits of pit pairs leading to a contiguous longitudinal tracheid

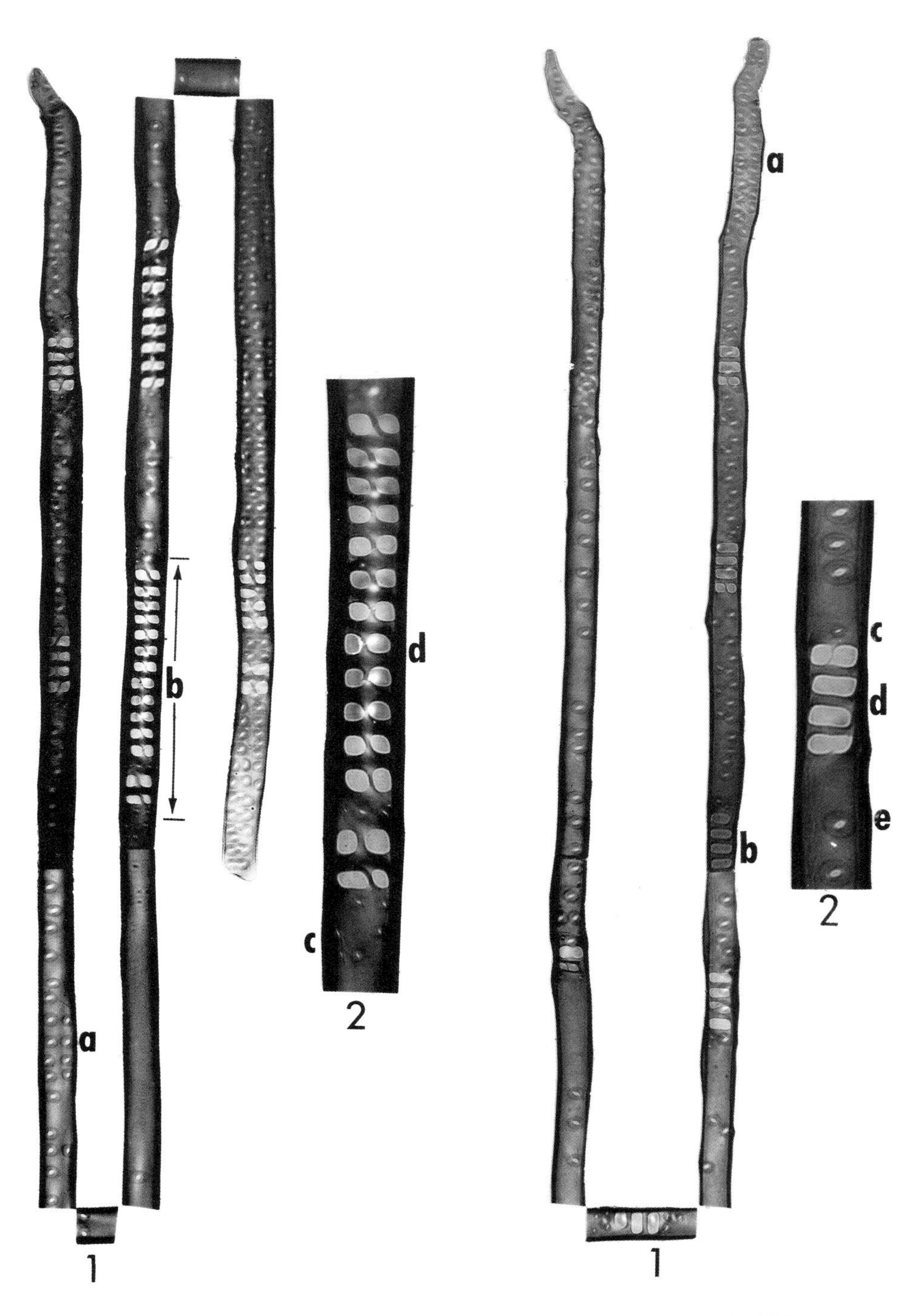

PLATE 2

RAY CELL TYPES IN HARD PINES

Ray Tracheids

1. Dentate ray tracheids. *Pinus ponderosa* Laws., 285X
    a. Dentations
2. Dentate ray tracheids. *Pinus resinosa* Ait., 285X
3. Dentate ray tracheids. *Pinus sylvestris* L., 285X
    b. Bordered pits in back wall
4. Dentate ray tracheids. *Pinus echinata* Mill., 285X

Ray Parenchyma

5. Ray parenchyma. *Pinus ponderosa* Laws., 285X
6. Ray parenchyma. *Pinus sylvestris* L., 285X

PLATE 3

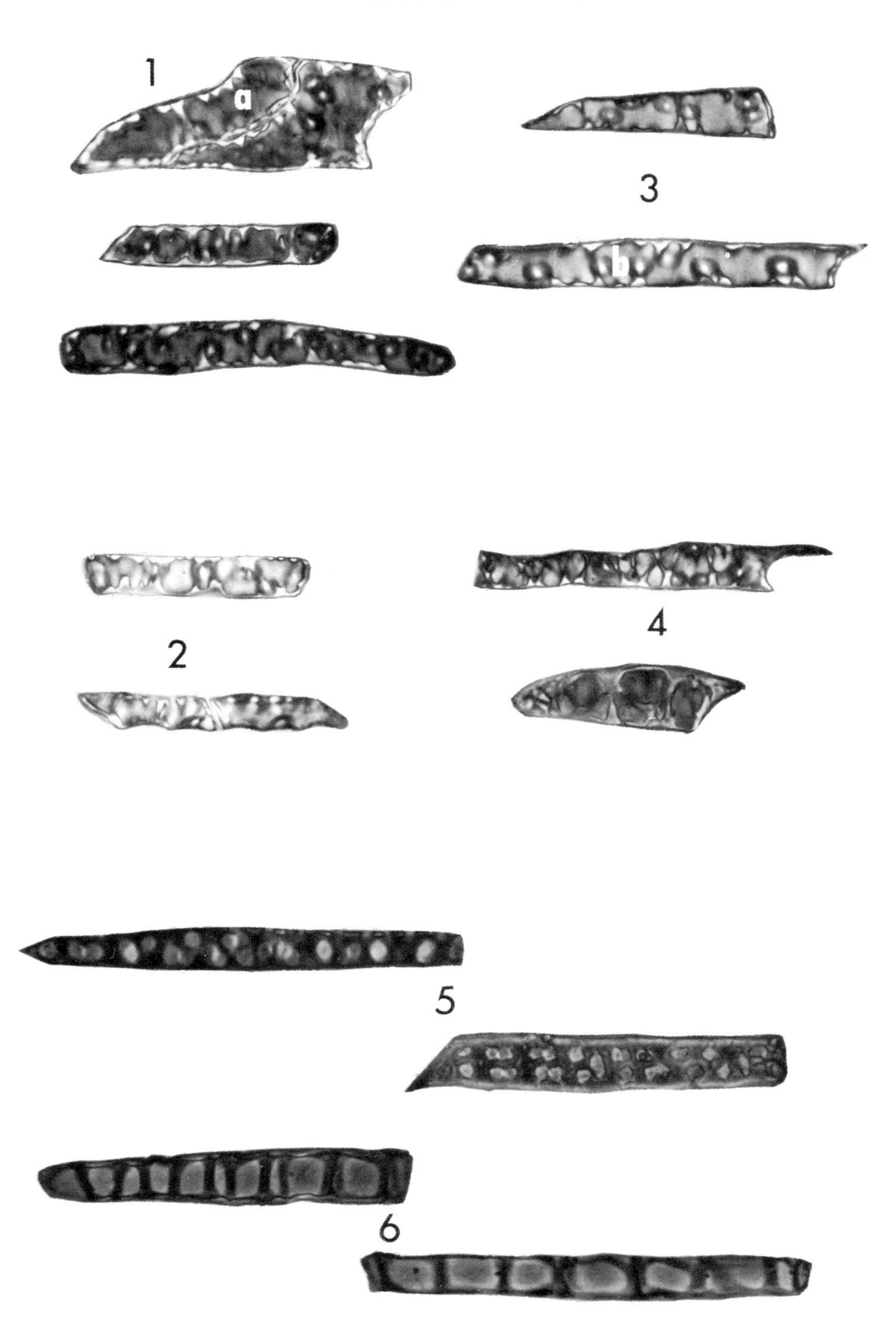
1
a
3
b
2
4
5
6

SCOTCH PINE, SCOTS PINE

Pinus sylvestris L.

1. Springwood longitudinal tracheid, 95X, radial view
   - *a.* Intertracheid pits: bordered pits of pit pairs leading to a contiguous longitudinal tracheid
   - *b.* Ray-contact area
2. Portion of a longitudinal tracheid, 185X
   - *c.* Ray-tracheid pits: small, bordered pits of pit pairs leading to a ray tracheid
   - *d.* Ray-parenchyma pits: large, window-like pits of pit pairs leading to ray parenchyma
   - *e.* Intertracheid pits: bordered pits of pit pairs leading to a contiguous longitudinal tracheid

RED PINE, NORWAY PINE

Pinus resinosa Ait.

1. Springwood longitudinal tracheid, 95X, radial view
   - *a.* Intertracheid pits: bordered pits of pit pairs leading to a contiguous longitudinal tracheid
   - *b.* Ray-contact area
2. Portion of a longitudinal tracheid, 185X
   - *c.* Ray-tracheid pits: small, bordered pits of pit pairs leading to a ray tracheid
   - *d.* Ray-parenchyma pits: large, window-like pits of pit pairs leading to ray parenchyma
   - *e.* Intertracheid pits: bordered pits of pit pairs leading to a contiguous longitudinal tracheid

PLATE 4

SCOTCH PINE

RED PINE

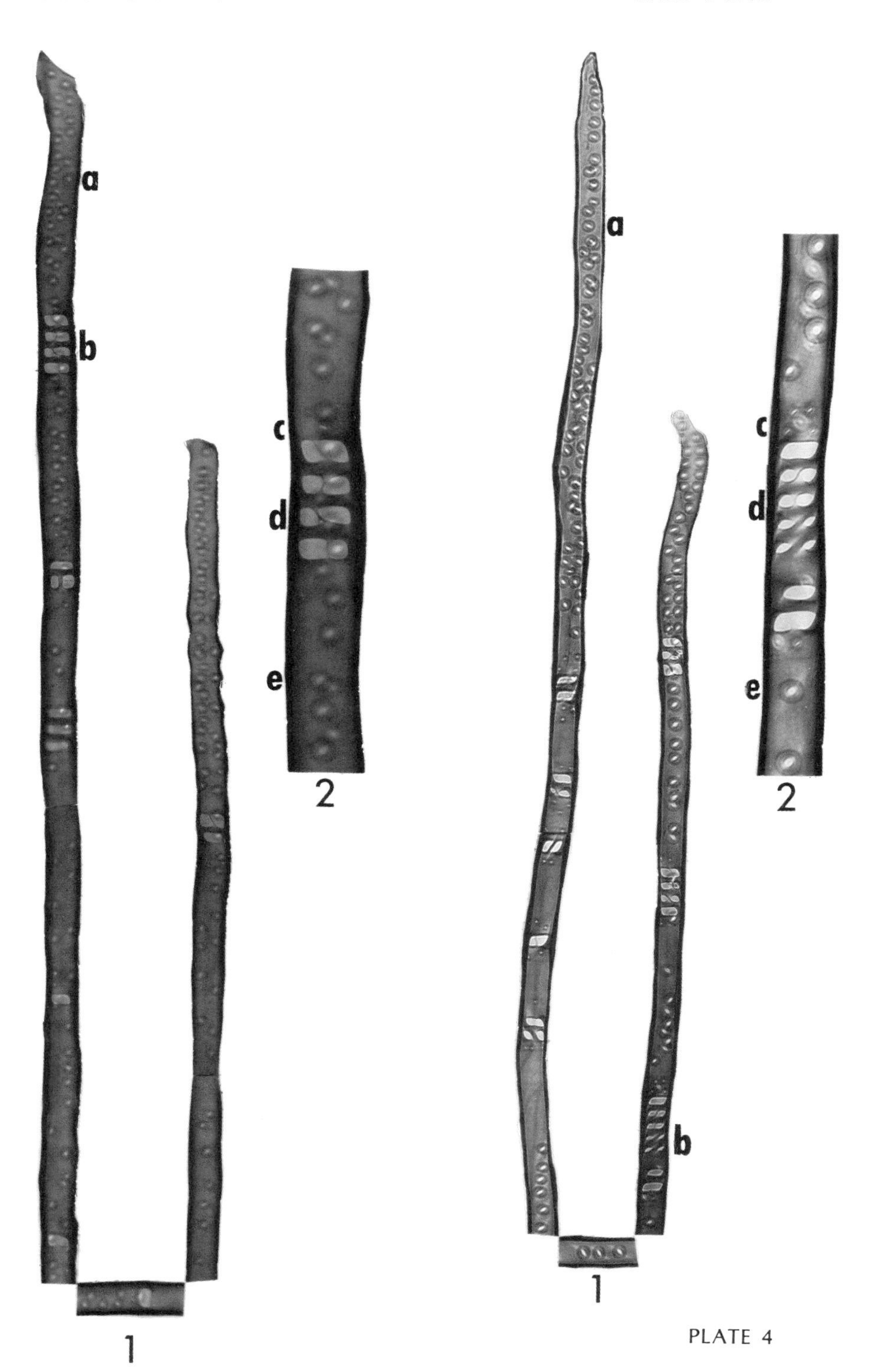

PLATE 4

MONTEREY PINE

Pinus radiata D. Don, *Pinus insignis* Dougl.

1. Springwood longitudinal tracheid, 95X, radial view
   - *a.* Intertracheid pits: bordered pits of pit pairs leading to a contiguous longitudinal tracheid
   - *b.* Ray-contact area

2. Portion of two longitudinal tracheids, 185X
   - *c.* Ray-tracheid pits: small, bordered pits of pit pairs leading to a ray tracheid
   - *d.* Ray-parenchyma pits: large, varishaped pinoid pits of pit pairs leading to ray parenchyma

3. Portion of a longitudinal tracheid, 185X, showing pit variation

PLATE 5

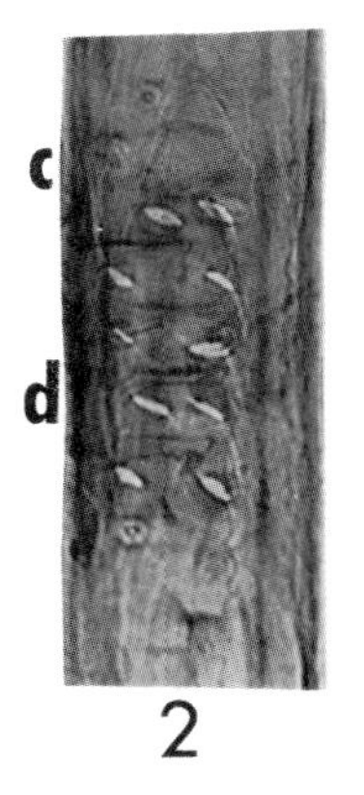

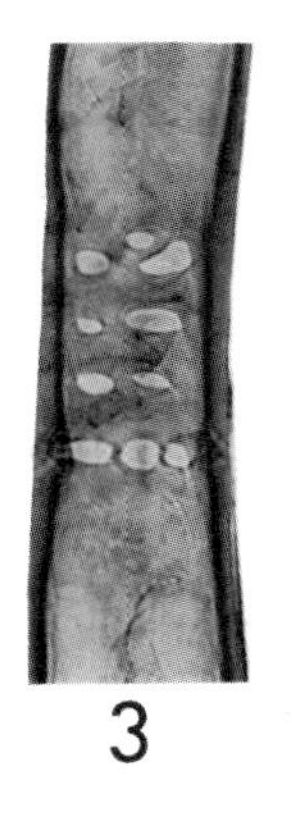

PLATE 5

SHORTLEAF PINE

Pinus echinata Mill.

1. Springwood longitudinal tracheid, 95X, radial view
   - *a.* Intertracheid pits: bordered pits of pit pairs leading to a contiguous longitudinal tracheid
   - *b.* Ray-contact area
2. Portion of two longitudinal tracheid, 185X
   - *c.* Ray-tracheid pits: small, bordered pits of pit pairs leading to a ray tracheid
   - *d.* Ray-parenchyma pits: large, varishaped pinoid pits of pit pairs leading to ray parenchyma
   - *e.* Intertracheid pits: bordered pits of pit pairs leading to a contiguous longitudinal tracheid

LOBLOLLY PINE

Pinus taeda L.

1. Springwood longitudinal tracheid, 95X, radial view
   - *a.* Intertracheid pits: bordered pits of pit pairs leading to a contiguous longitudinal tracheid
   - *b.* Ray-contact area
2. Portions of two contiguous longitudinal tracheids, 185X
   - *c.* Ray-tracheid pits: small, bordered pits of pit pairs leading to a ray tracheid
   - *d.* Ray-parenchyma pits: large, varishaped pinoid pits of pit pairs leading to ray parenchyma
   - *e.* Intertracheid pits: bordered pits of pit pairs leading to a contiguous longitudinal tracheid

SHORTLEAF PINE

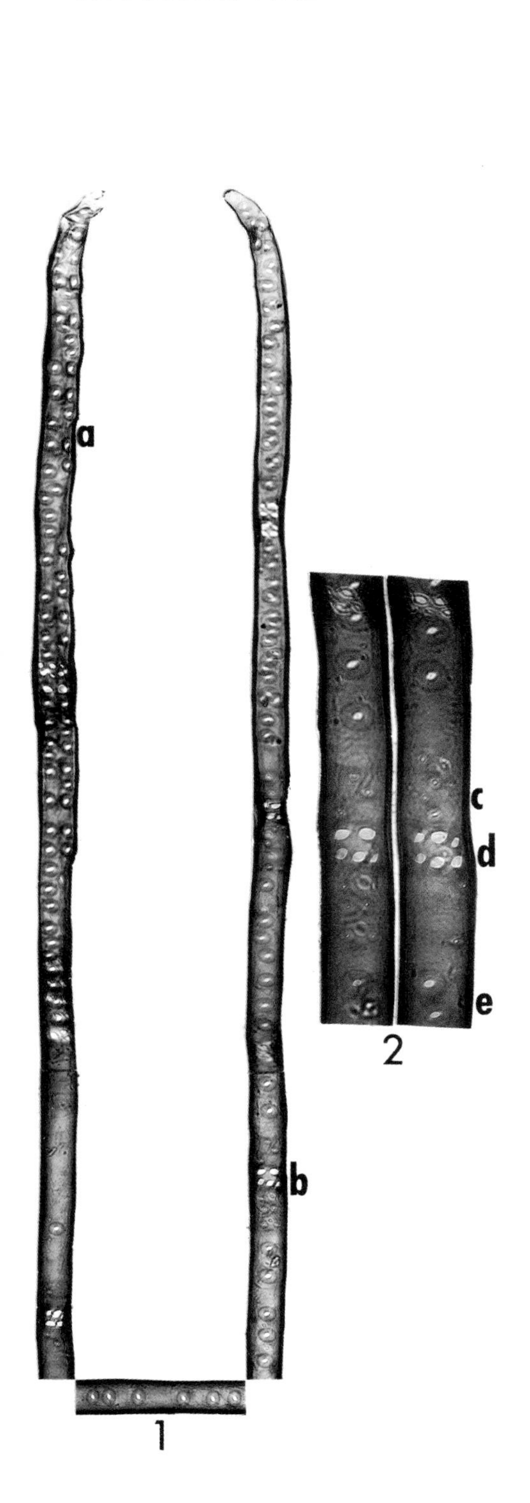

LOBLOLLY PINE

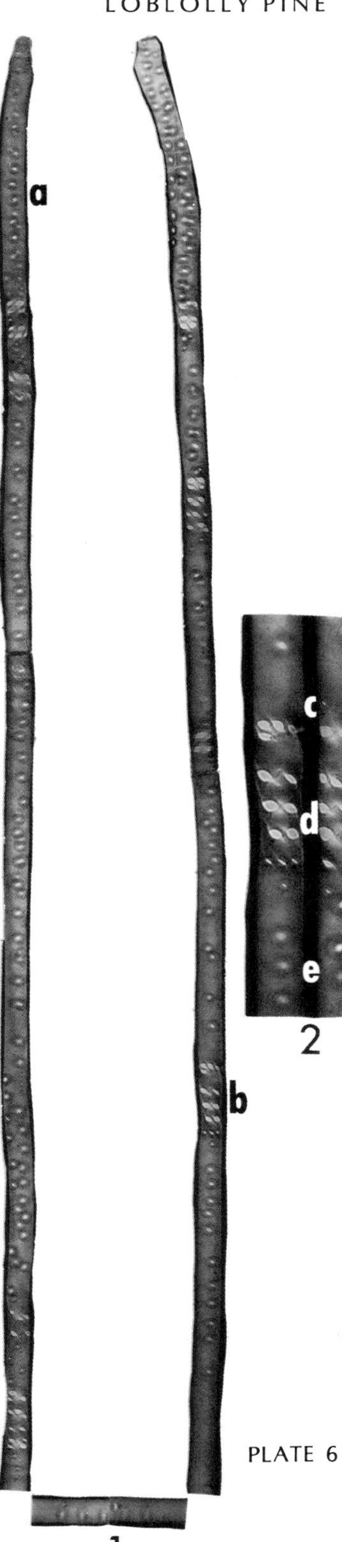

PLATE 6

LONGLEAF PINE

Pinus palustris Mill.

1. Springwood longitudinal tracheid, 95X, radial view
   - *a.* Intertracheid pits: bordered pits of pit pairs leading to a contiguous longitudinal tracheid
   - *b.* Ray-contact area
2. Portion of a longitudinal tracheid, 185X
   - *c.* Ray-tracheid pits: small, bordered pits of pit pairs leading to a ray tracheid
   - *d.* Ray-parenchyma pits: large, varishaped pinoid pits of pit pairs leading to ray parenchyma
   - *e.* Intertracheid pits: bordered pits of pit pairs leading to a contiguous longitudinal tracheid
3. Portion of a longitudinal tracheid, 185X, showing pit variation

SLASH PINE

Pinus elliottii Engelm.

1. Springwood longitudinal tracheid, 95X, radial view
   - *a.* Intertracheid pits: bordered pits of pit pairs leading to a contiguous longitudinal tracheid
   - *b.* Ray-contact area
2. Portion of a longitudinal tracheid, 185X
   - *c.* Ray-tracheid pits: small, bordered pits of pit pairs leading to a ray tracheid
   - *d.* Ray-parenchyma pits: large, varishaped pinoid pits of pit pairs leading to ray parenchyma
3. Portion of a longitudinal tracheid, 185X, showing pit variation

LONGLEAF PINE

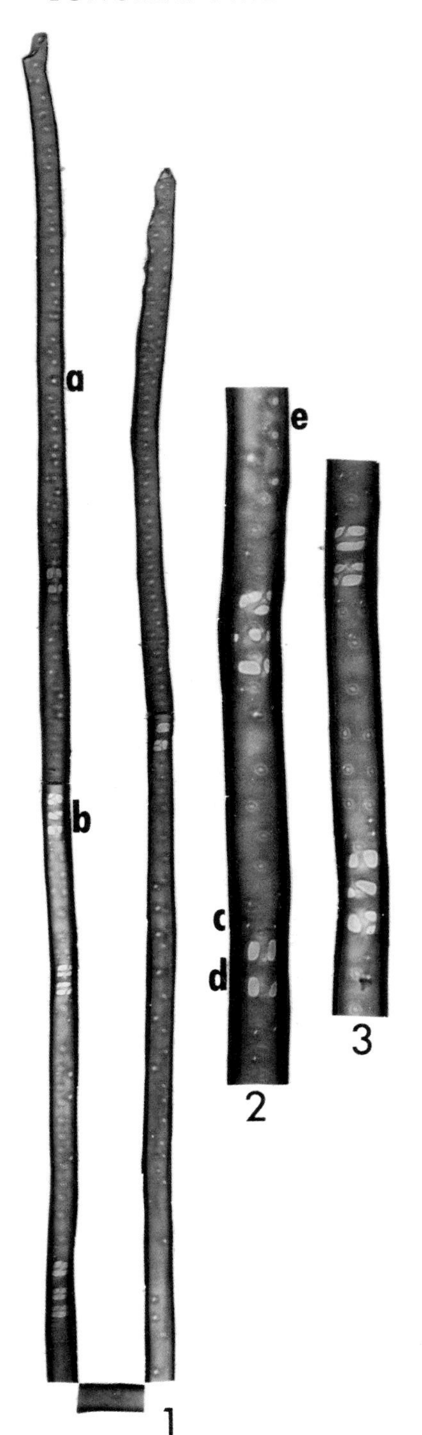

SLASH PINE

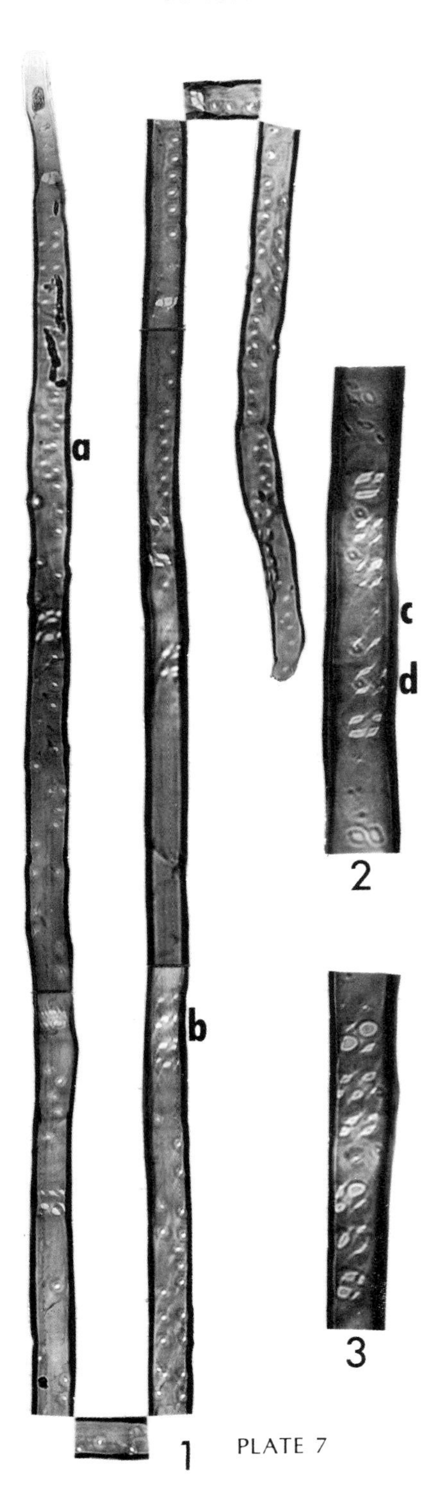

PLATE 7

PONDEROSA PINE, WESTERN YELLOW PINE

Pinus ponderosa Laws.

1. Springwood longitudinal tracheid, 95X, radial view
   - *a.* Intertracheid pits: bordered pits of pit pairs leading to a contiguous longitudinal tracheid
   - *b.* Ray-contact area
2. Portions of two contiguous longitudinal tracheids, 95X
   - *c.* Ray-tracheid pits: small, bordered pits of pit pairs leading to a ray tracheid
   - *d.* Ray-parenchyma pits: large, varishaped pinoid pits of pit pairs leading to ray parenchyma
   - *e.* Intertracheid pits: bordered pits of pit pairs leading to a contiguous longitudinal tracheid

JACK PINE

Pinus banksiana Lamb.

1. Springwood longitudinal tracheid, 95X, radial view
   - *a.* Intertracheid pits: bordered pits of pit pairs leading to a contiguous longitudinal tracheid
   - *b.* Ray-contact area
2. Portion of a longitudinal tracheid, 185X
   - *c.* Ray-tracheid pits: small, bordered pits of pit pairs leading to a ray tracheid
   - *d.* Ray-parenchyma pits: large, varishaped pinoid pits of pit pairs leading to ray parenchyma
   - *e.* Intertracheid pits: bordered pits of pit pairs leading to a contiguous longitudinal tracheid
3. Portion of a longitudinal tracheid, 185X, showing pit variation

PONDEROSA PINE

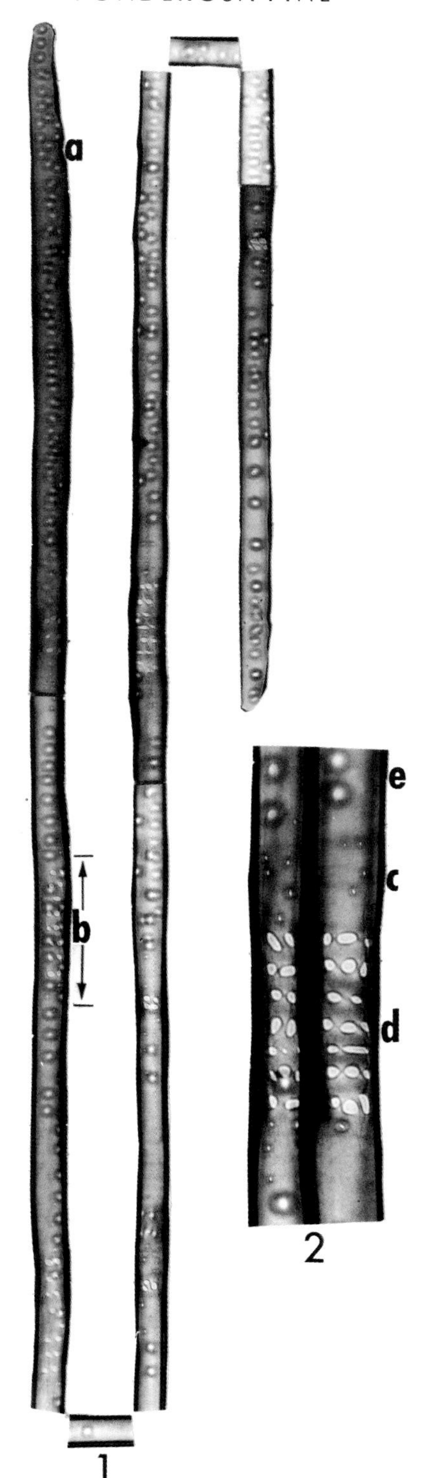

JACK PINE

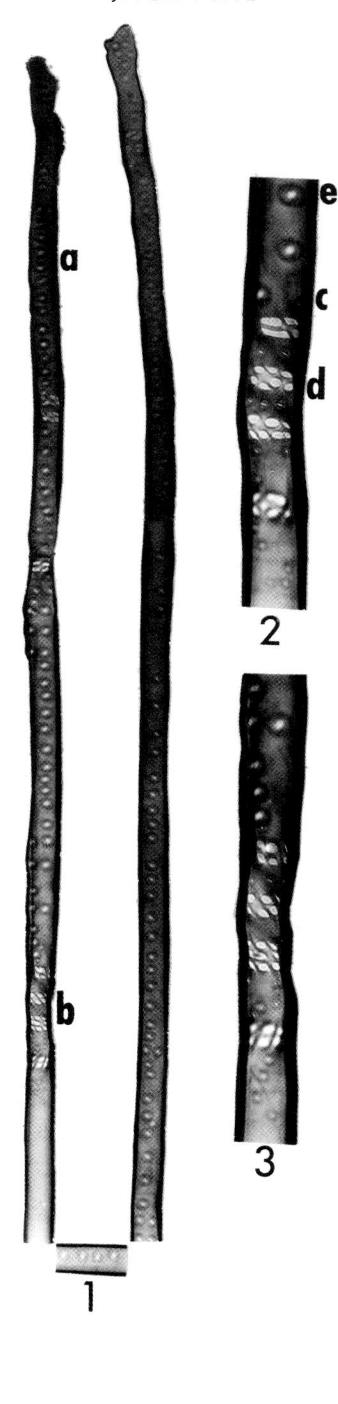

PLATE 8

PITCH PINE

Pinus rigida Mill.

1. Springwood longitudinal tracheid, 95X, radial view
   - *a.* Intertracheid pits: bordered pits of pit pairs leading to a contiguous longitudinal tracheid
   - *b.* Ray-contact area
2. Portion of a longitudinal tracheid, 185X
   - *c.* Ray-tracheid pits: small, bordered pits of pit pairs leading to a ray tracheid
   - *d.* Ray-parenchyma pits: large, varishaped pinoid pits of pit pairs leading to ray parenchyma
   - *e.* Intertracheid pits: bordered pits of pit pairs leading to a contiguous longitudinal tracheid
3. Portion of a longitudinal tracheid, 185X, showing pit variation

WESTERN LARCH

Laris occidentalis Nutt.

1. Springwood longitudinal tracheid, 95X, radial view
   - *a.* Intertracheid pits: bordered pits of pit pairs leading to a contiguous longitudinal tracheid
   - *b.* Ray-contact area
2. Portion of a longitudinal tracheid, 185X
   - *c.* Ray-tracheid pits: small, bordered pits of pit pairs leading to a ray tracheid
   - *d.* Ray-parenchyma pits: small, slitlike piciform pits of pit pairs leading to ray parenchyma
   - *e.* Intertracheid pits: bordered pits of pit pairs leading to a contiguous longitudinal tracheid

PLATE 9

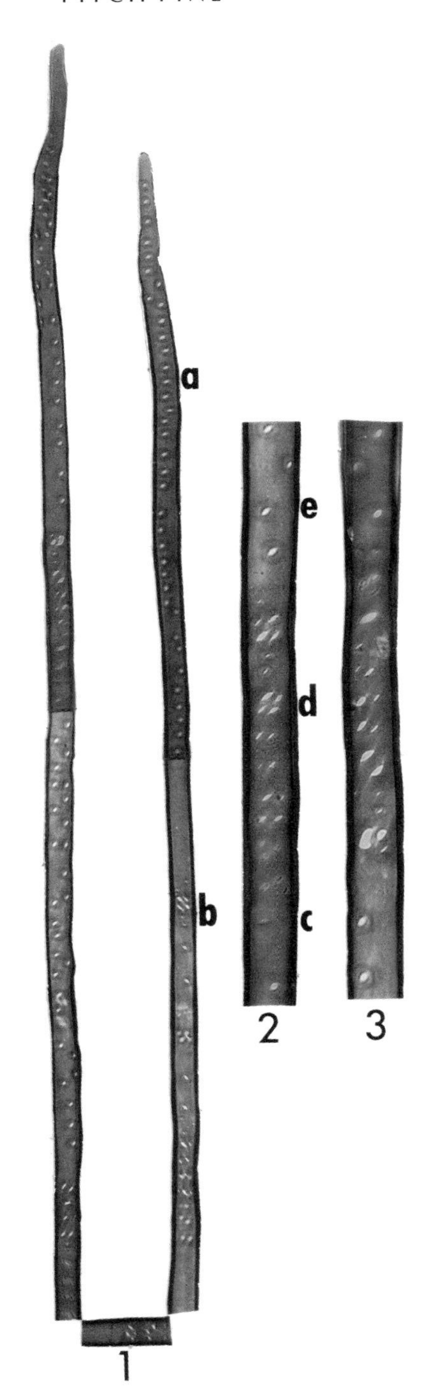
PITCH PINE
a
b
e
d
c
1
2
3

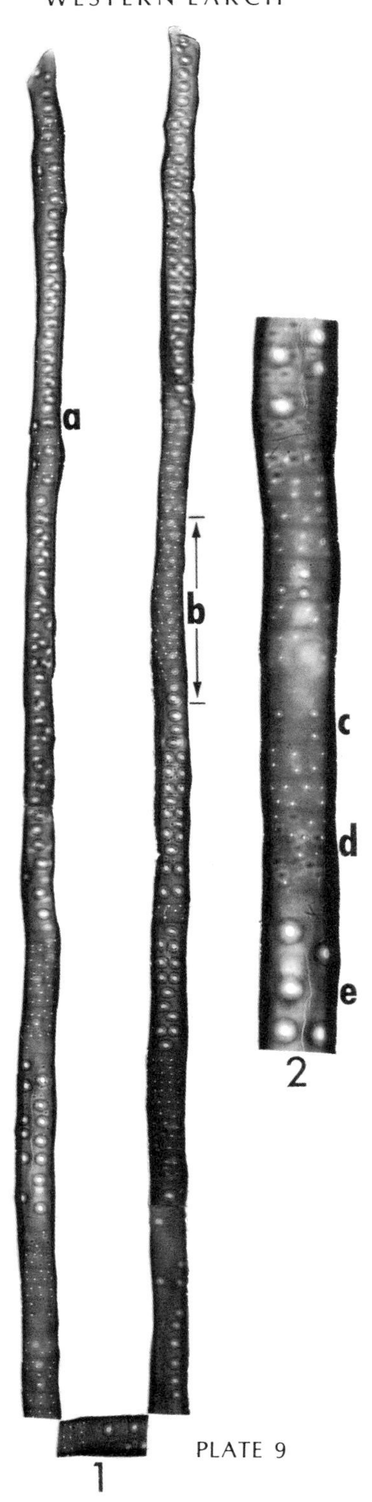
WESTERN LARCH
a
b
c
d
e
1
2

EUROPEAN LARCH

Larix decidua Mill.

1. Springwood longitudinal tracheid, 95X, radial view
   - *a.* Intertracheid pits: bordered pits of pit pairs leading to a contiguous longitudinal tracheid
   - *b.* Ray-contact area
2. Portion of a longitudinal tracheid, 185X
   - *c.* Ray-tracheid pits: small, bordered pits of pit pairs leading to a ray tracheid
   - *d.* Ray-parenchyma pits: small, slitlike piciform pits of pit pairs leading to ray parenchyma
   - *e.* Intertracheid pits: bordered pits of pit pairs leading to a contiguous longitudinal tracheid

EASTERN LARCH, TAMARACK

Larix laricina (DuRoi) K. Koch

1. Springwood longitudinal tracheid, 95X, radial view
   - *a.* Intertracheid pits: bordered pits of pit pairs leading to a contiguous longitudinal tracheid
   - *b.* Ray-contact areas
2. Portion of a longitudinal tracheid, 185X
   - *c.* Ray-tracheid pits: small, bordered pits of pit pairs leading to a ray tracheid
   - *d.* Ray-parenchyma pits: small, slitlike piciform pits of pit pairs leading to ray parenchyma
   - *e.* Intertracheid pits: bordered pits of pit pairs leading to a contiguous longitudinal tracheid

EUROPEAN LARCH

EASTERN LARCH

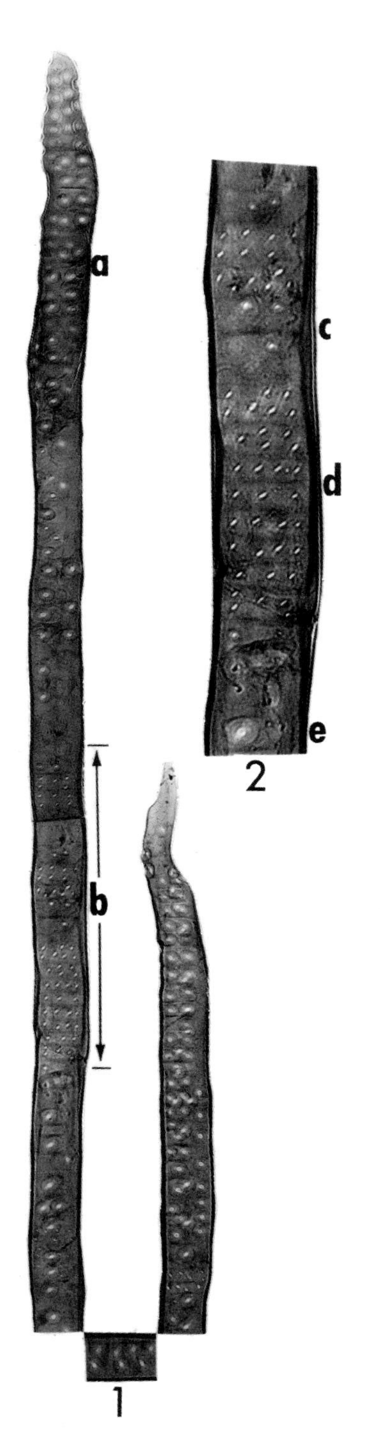

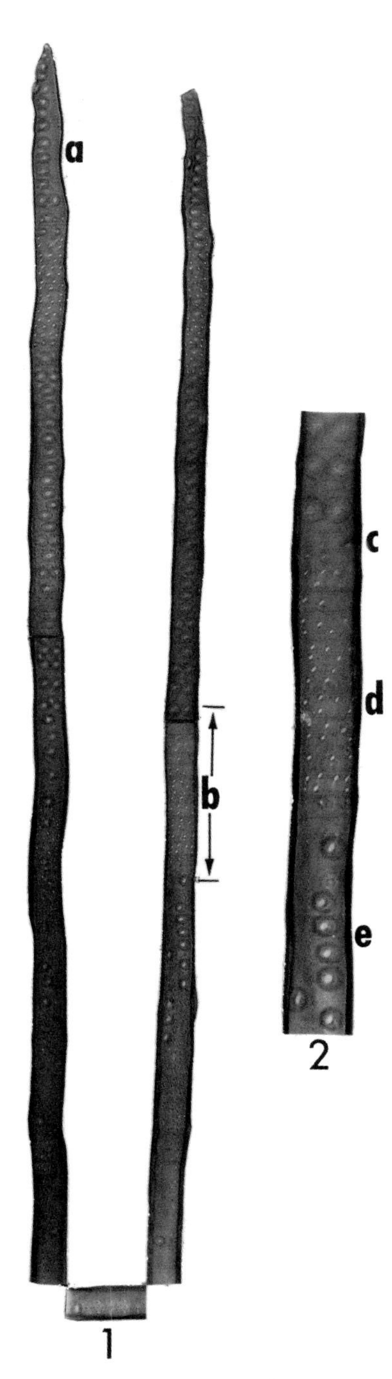

RED SPRUCE
Picea rubens Sarg.

1. Springwood longitudinal tracheid, 95X, radial view
   - *a.* Intertracheid pits: bordered pits of pit pairs leading to a contiguous longitudinal tracheid
   - *b.* Ray-contact area
2. Portion of a longitudinal tracheid, 185X, showing pit variation
3. Portion of a longitudinal tracheid, 185X
   - *c.* Ray-tracheid pits: small, bordered pits of pit pairs leading to a ray tracheid
   - *d.* Ray-parenchyma pits: small, slitlike piciform pits of pit pairs leading to ray parenchyma
   - e. Intertracheid pits: bordered pits of pit pairs leading to a contiguous longitudinal tracheid

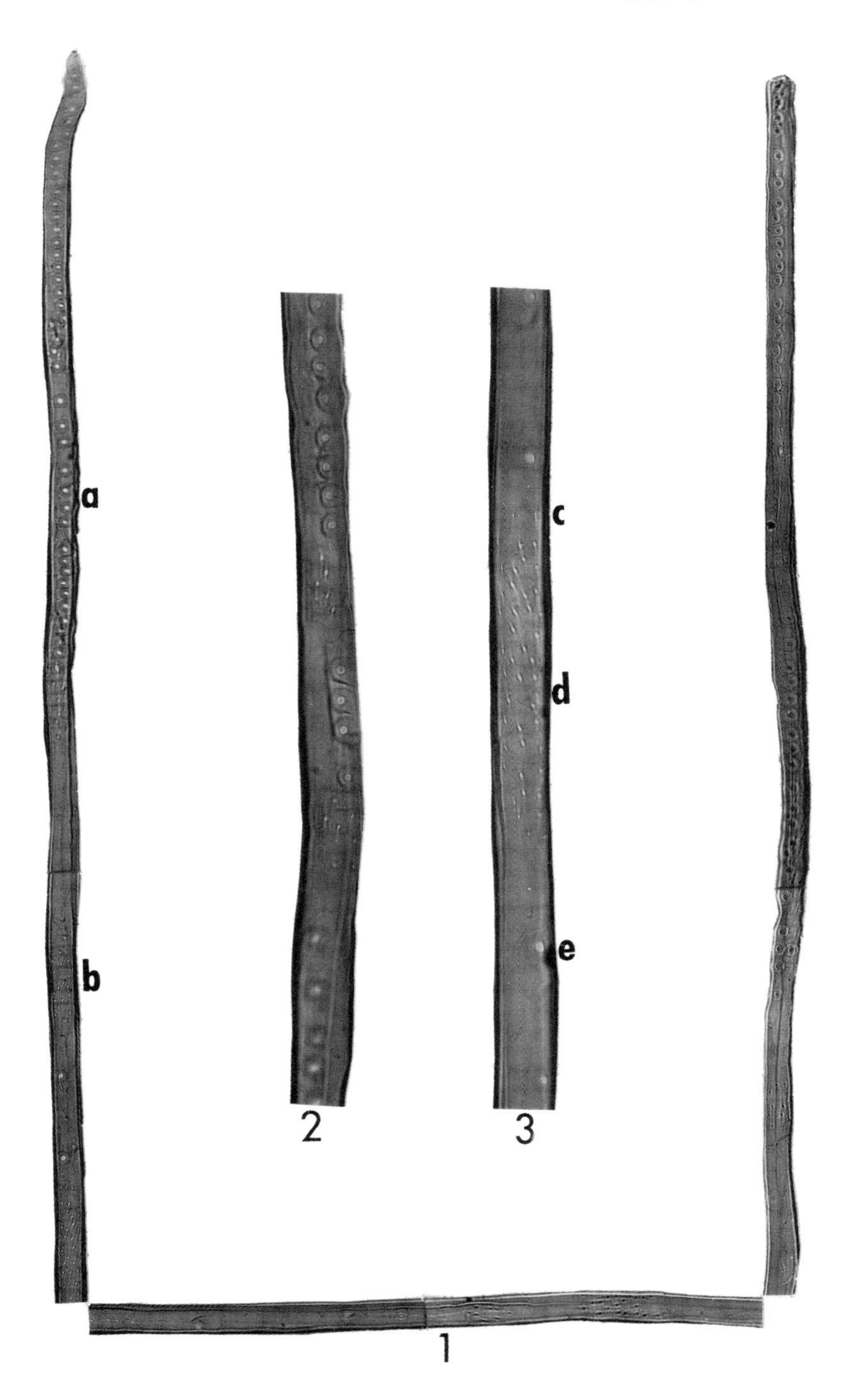
a
b
c
d
e
2
3
1

WHITE SPRUCE
Picea glauca (Moench) Voss.

1. Springwood longitudinal tracheid, 95X, radial view
   - *a.* Intertracheid pits: bordered pits of pit pairs leading to a contiguous longitudinal tracheid
   - *b.* Ray-contact area
2. Portion of a longitudinal tracheid, 185X
   - *c.* Ray-tracheid pits: small, bordered pits of pit pairs leading to a ray tracheid
   - *d.* Ray-parenchyma pits: small, slitlike piciform pits of pit pairs leading to ray parenchyma
   - *e.* Intertracheid pits: bordered pits of pit pairs leading to a contiguous longitudinal tracheid

BLACK SPRUCE
Picea mariana (Mill.) B.S.P.

1. Springwood longitudinal tracheid, 95X, radial view
   - *a.* Intertracheid pits: bordered pits of pit pairs leading to a contiguous longitudinal tracheid
   - *b.* Ray-contact area
2. Portion of a longitudinal tracheid, 185X
   - *c.* Ray-tracheid pits: small, bordered pits of pit pairs leading to a ray tracheid
   - *d.* Ray-parenchyma pits: small, slitlike piciform pits of pit pairs leading to ray parenchyma
   - *e.* Intertracheid pits: bordered pits of pit pairs leading to a contiguous longitudinal tracheid

WHITE SPRUCE

BLACK SPRUCE

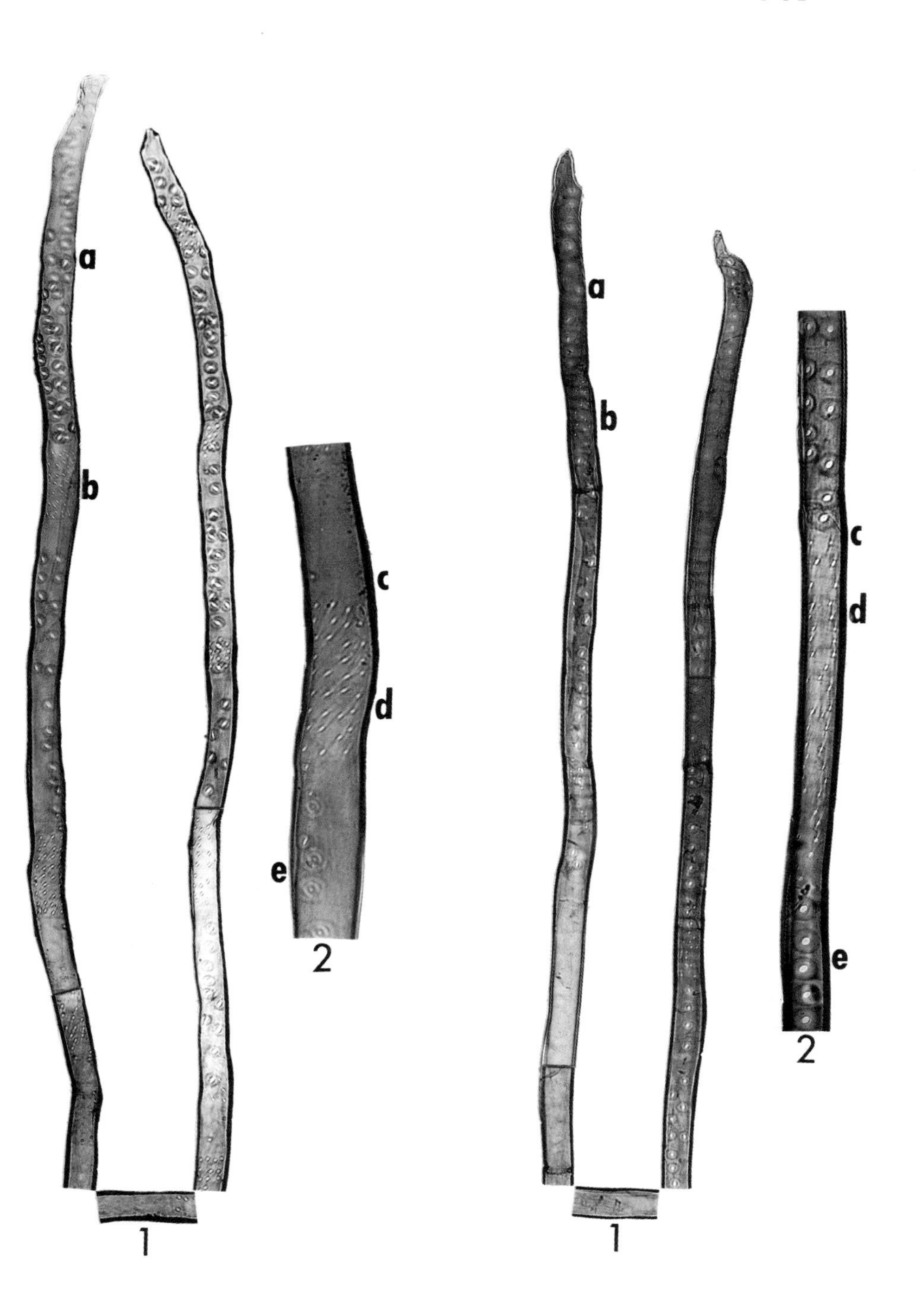

PLATE 12

SITKA SPRUCE

Picea sitchensis (Bong.) Carr.

1. Springwood longitudinal tracheid, 95X, radial view
   - *a.* Intertracheid pits: bordered pits of pit pairs leading to a contiguous longitudinal tracheid
   - *b.* Ray-contact area
2. Portion of a longitudinal tracheid, 185X
   - *c.* Ray-tracheid pits: small, bordered pits of pit pairs leading to a ray tracheid
   - *d.* Ray-parenchyma pits: small, slitlike piciform pits of pit pairs leading to ray parenchyma
   - e. Intertracheid pit: bordered pit of pit pairs leading to a contiguous longitudinal tracheid

PLATE 13

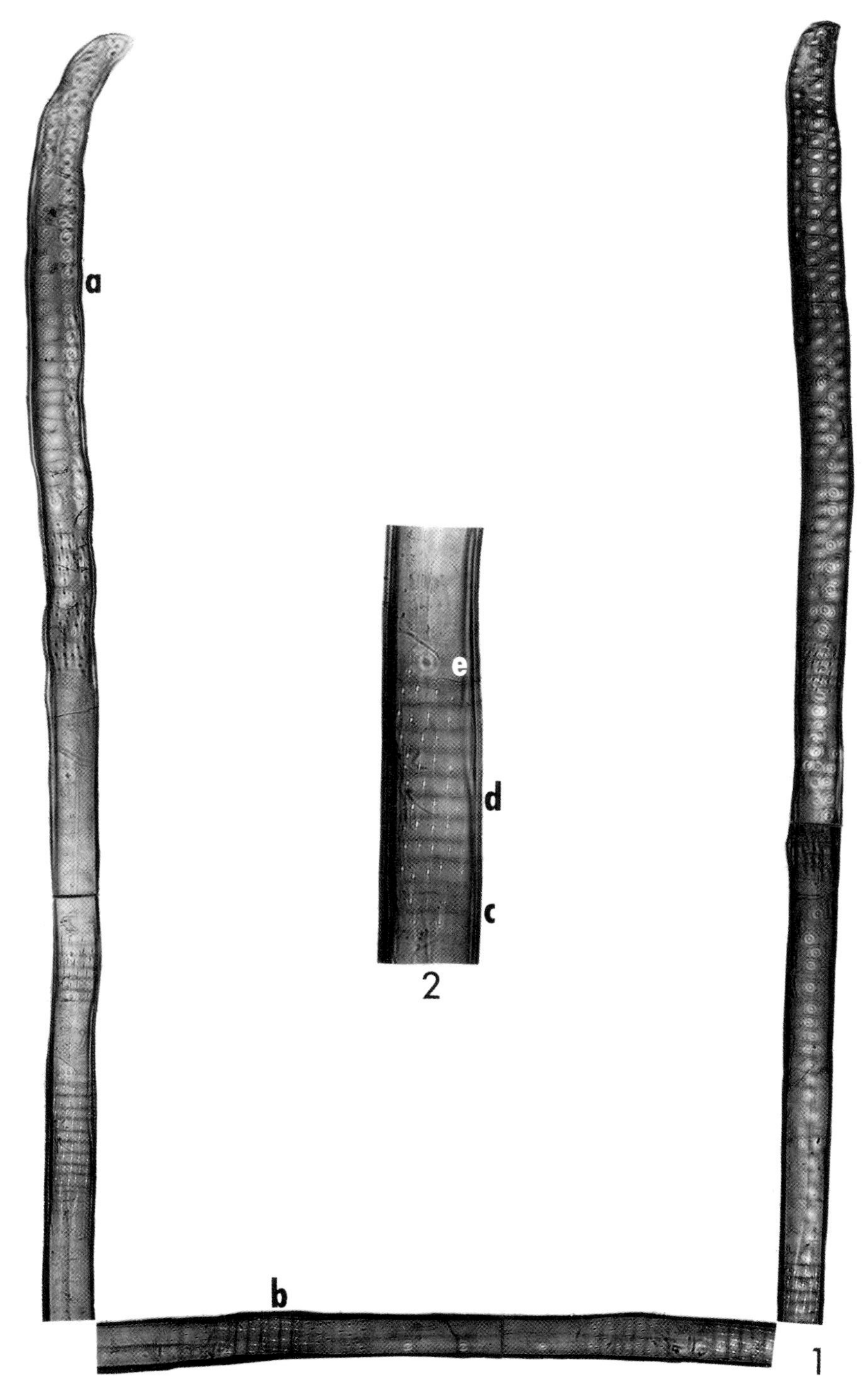

PLATE 13

ENGELMANN SPRUCE

Picea engelmannii Parry

Springwood longitudinal tracheid, 185X, radial view

a. Intertracheid pits: bordered pits of pit pairs leading to a contiguous longitudinal tracheid

b. Ray-contact area

c. Ray-tracheid pits: small, bordered pits of pit pairs leading to a ray tracheid

d. Ray-parenchyma pits: small, slitlike piciform pits of pit pairs leading to ray parenchyma

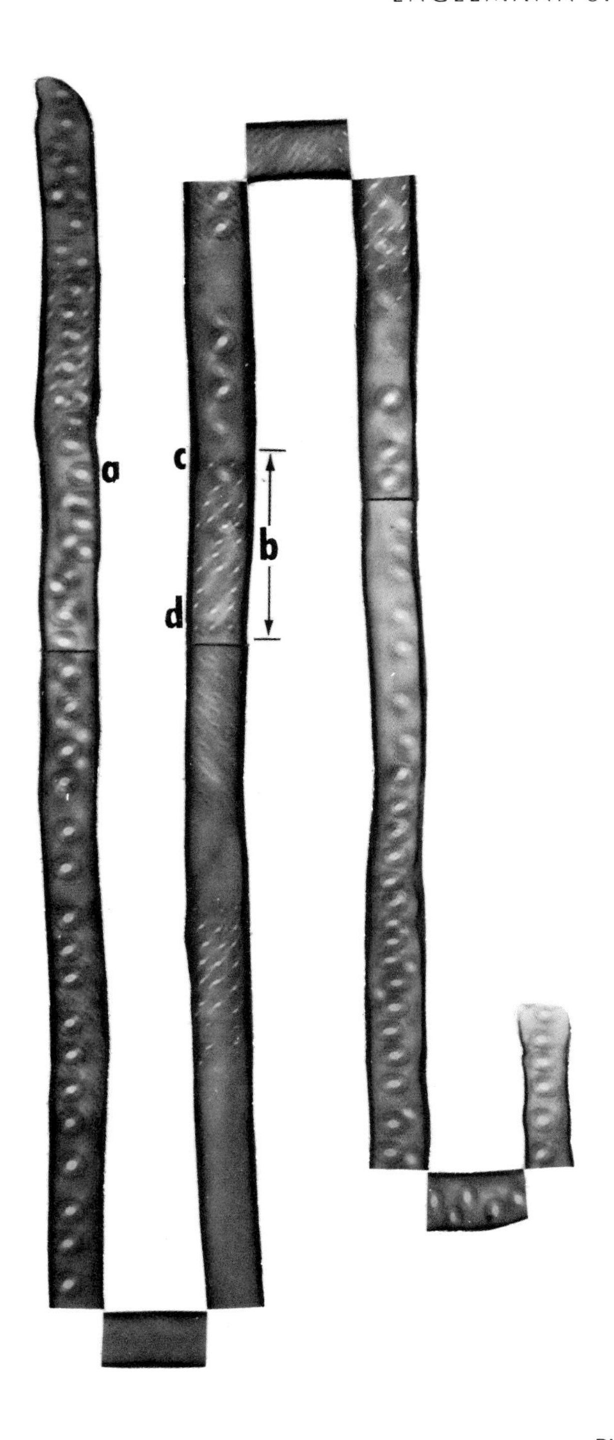
a
c
b
d

NORWAY SPRUCE

Picea abies (L.) Karst., *Picea excelsa* Link

Springwood longitudinal tracheid, 185X, radial view

a. Intertracheid pits: bordered pits of pit pairs leading to a contiguous longitudinal tracheid

b. Ray-contact area

c. Ray-tracheid pits: small, bordered pits of pit pairs leading to a ray tracheid

d. Ray-parenchyma pits: small, slitlike piciform pits of pit pairs leading to ray parenchyma

PLATE 15

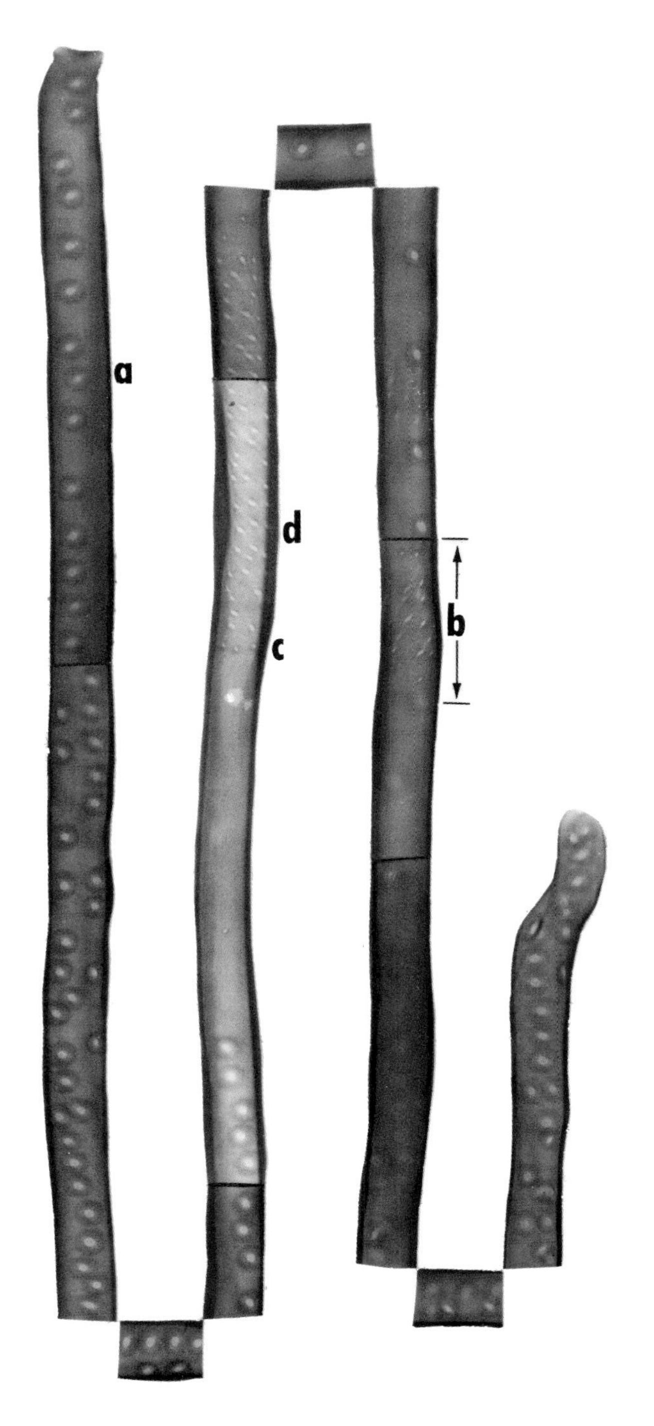
a
d
c
b

DOUGLAS-FIR

Pseudotsuga menziesii (Mirb.) Franco

1. Springwood longitudinal tracheid, 95X, radial view
   - *a.* Intertracheid pits: bordered pits of pit pairs leading to a contiguous longitudinal tracheid
   - *b.* Ray-contact area
2. Portion of a longitudinal tracheid, 185X
   - *c.* Ray-tracheid pits: small, bordered pits of pit pairs leading to a ray tracheid
   - *d.* Ray-parenchyma pits: small, slitlike piciform pits of pit pairs leading to ray parenchyma
   - *e.* Intertracheid pits: bordered pits of pit pairs leading to a contiguous longitudinal tracheid
   - *f.* Helical thickening: localized thickening of the secondary wall.

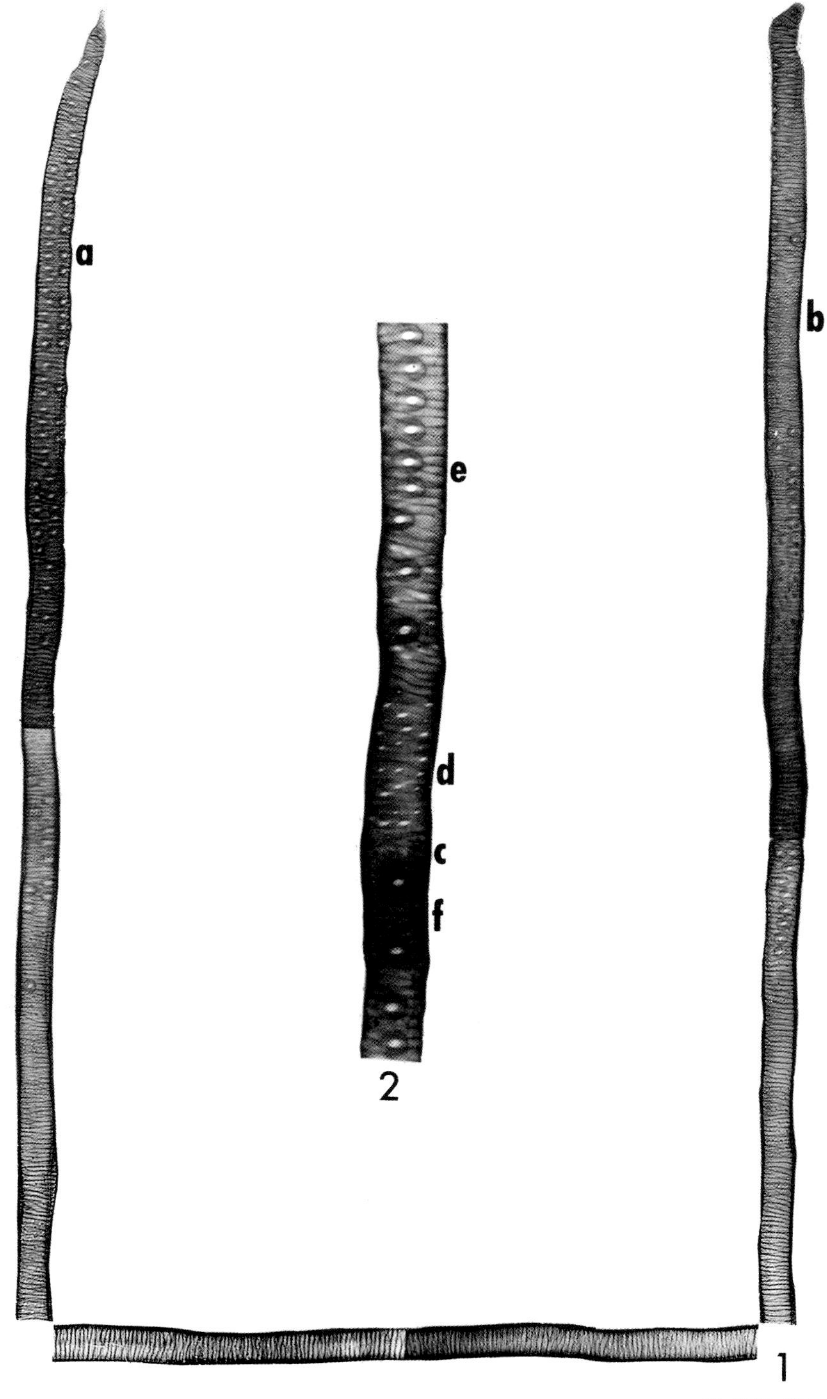

PLATE 16

EASTERN HEMLOCK

Tsuga canadensis (L.) Carr.

1. Springwood longitudinal tracheid, 95X, radial view
   - *a.* Intertracheid pits: bordered pits of pit pairs leading to a contiguous longitudinal tracheid
   - *b.* Ray-contact area
2. Portion of a longitudinal tracheid, 185X
   - *c.* Ray-tracheid pits: small, bordered pits of pit pairs leading to a ray tracheid
   - *d.* Ray-parenchyma pits: small, elliptical, uniformly sized pits of pit pairs leading to ray parenchyma
   - *e.* Intertracheid pits: bordered pits of pit pairs leading to a contiguous longitudinal tracheid

WESTERN HEMLOCK

Tsuga heterophylla (Raf.) Sarg.

1. Springwood longitudinal tracheid, 95X, radial view
   - *a.* Intertracheid pits: bordered pits of pit pairs leading to a contiguous longitudinal tracheid
   - *b.* Ray-contact area
2. Portion of a longitudinal tracheid, 185X
   - *c.* Ray-tracheid pits: small, bordered pits of pit pairs leading to a ray tracheid
   - *d.* Ray-parenchyma pits: small, elliptical, uniformly sized pits of pit pairs leading to ray parenchyma
   - *e.* Intertracheid pits: bordered pits of pit pairs leading to a contiguous longitudinal tracheid

EASTERN HEMLOCK

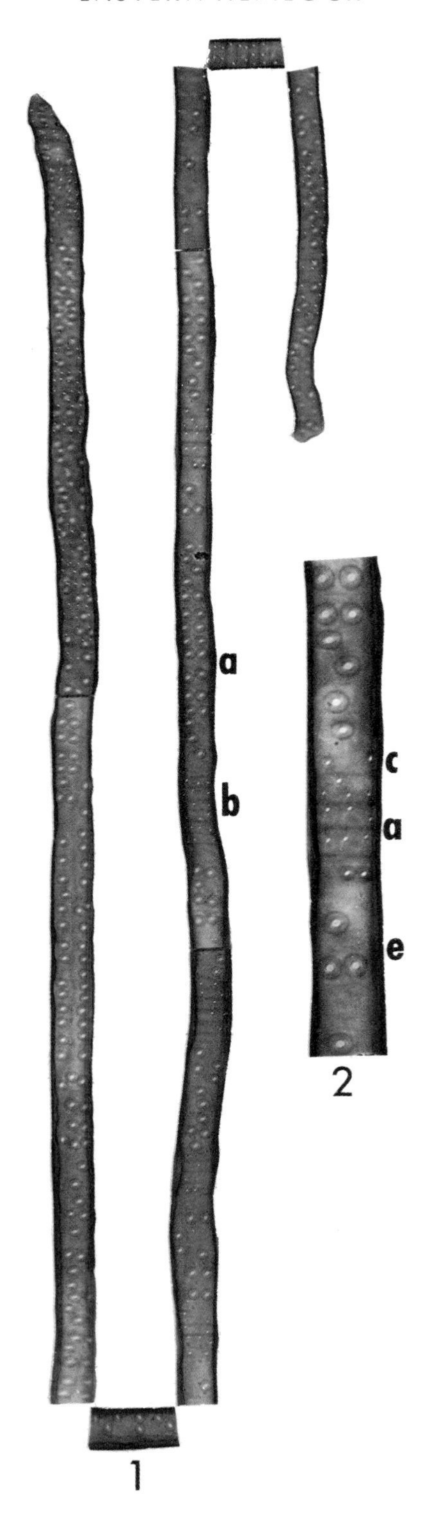

WESTERN HEMLOCK

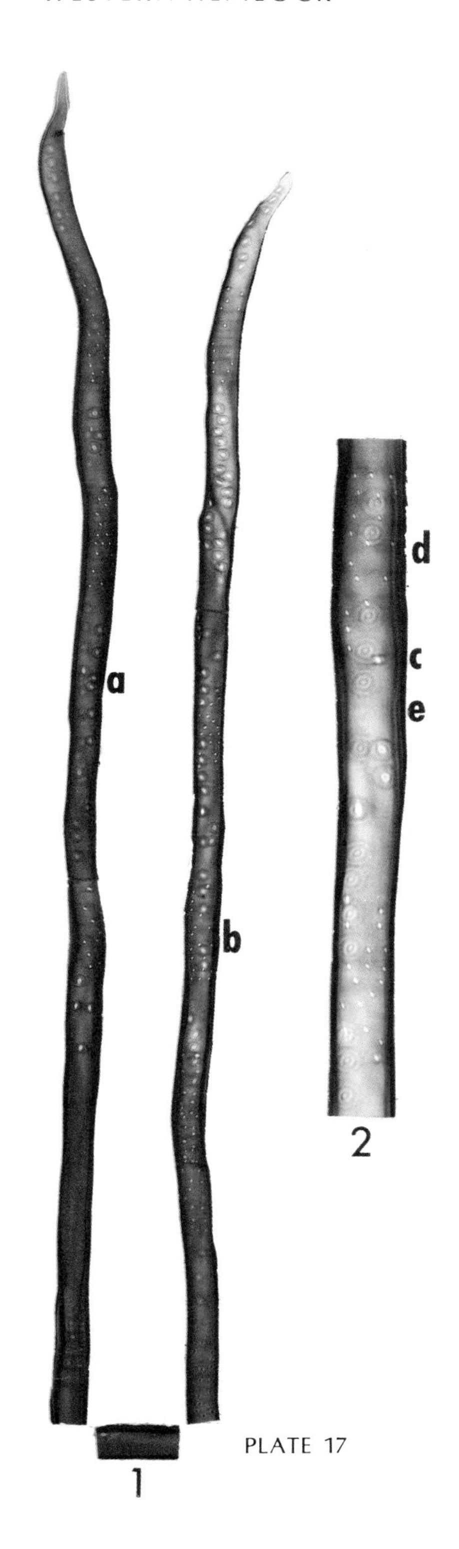

PLATE 17

SILVER FIR

Abies alba Mill.

1. Springwood longitudinal tracheid, 95X, radial view
   - *a.* Intertracheid pits: bordered pits of pit pairs leading to a contiguous longitudinal tracheid
   - *b.* Ray-contact area
2. Portion of a longitudinal tracheid, 185X
   - *c.* Ray-parenchyma pits: small, elliptical, uniformly sized pits of pit pairs leading to ray parenchyma
   - *d.* Intertracheid pits: bordered pits of pit pairs leading to a contiguous longitudinal tracheid

BALSAM FIR, EASTERN FIR

Abies balsamea (L.) Mill.

1. Springwood longitudinal tracheid, 95X, radial view
   - *a.* Intertracheid pits: bordered pits of pit pairs leading to a contiguous longitudinal tracheid
   - *b.* Ray-contact area
2. Portion of a longitudinal tracheid, 185X
   - *c.* Ray-contact area
     sized pits of pit pairs leading to ray parenchyma
   - *d.* Intertracheid pits: bordered pits of pit pairs leading to a contiguous longitudinal tracheid

SILVER FIR

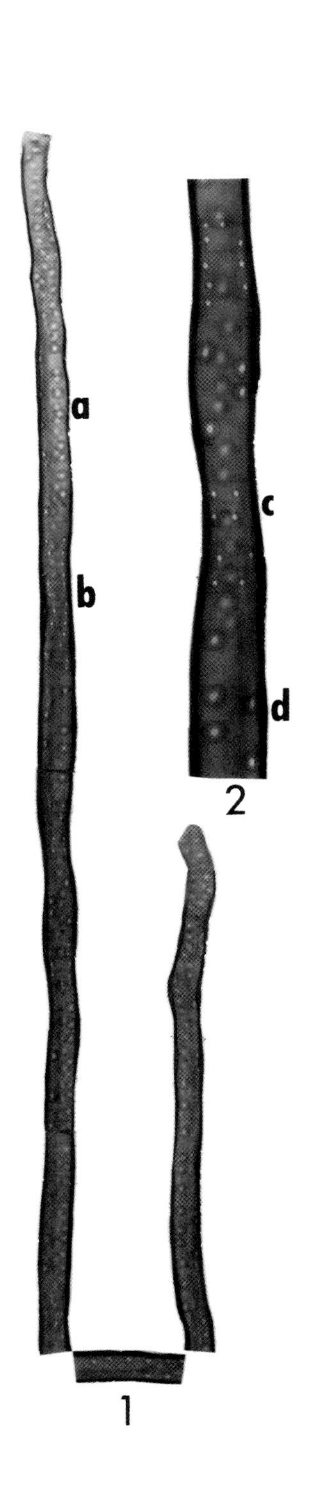

BALSAM FIR, EASTERN FIR

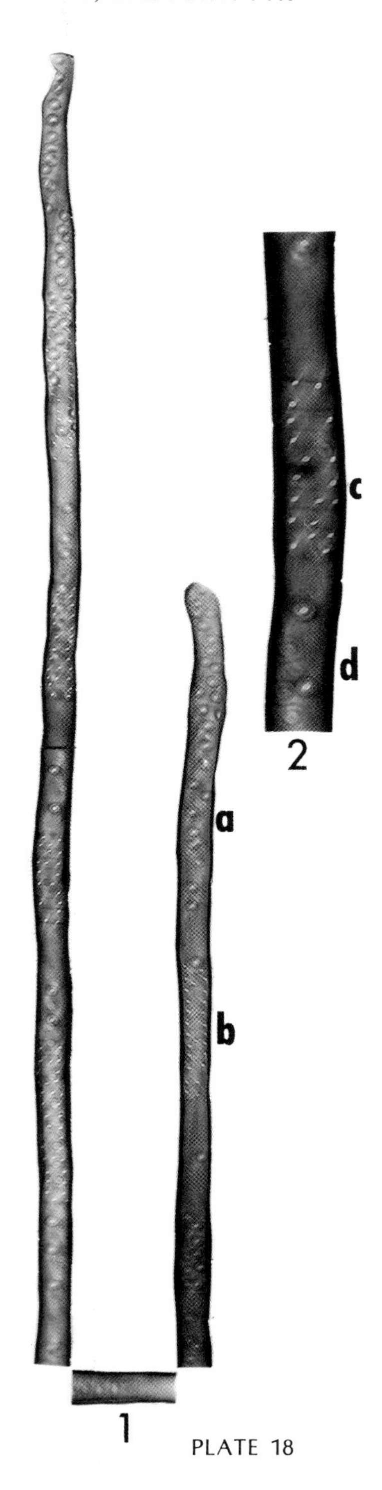

WHITE FIR

Abies concolor (Gord. and Glend.) Lindl.

1. Springwood longitudinal tracheid, 95X, radial view
   a. Intertracheid pits: bordered pits of pit pairs leading to a contiguous longitudinal tracheid
   b. Ray-contact area
2. Portion of a longitudinal tracheid, 185X
   c. Ray-parenchyma pits: small, elliptical, uniformly sized pits of pit pairs leading to ray parenchyma
   d. Intertracheid pits: bordered pits of pit pairs leading to a contiguous longitudinal tracheid
3. Portion of a longitudinal tracheid, 185X, showing pit variation

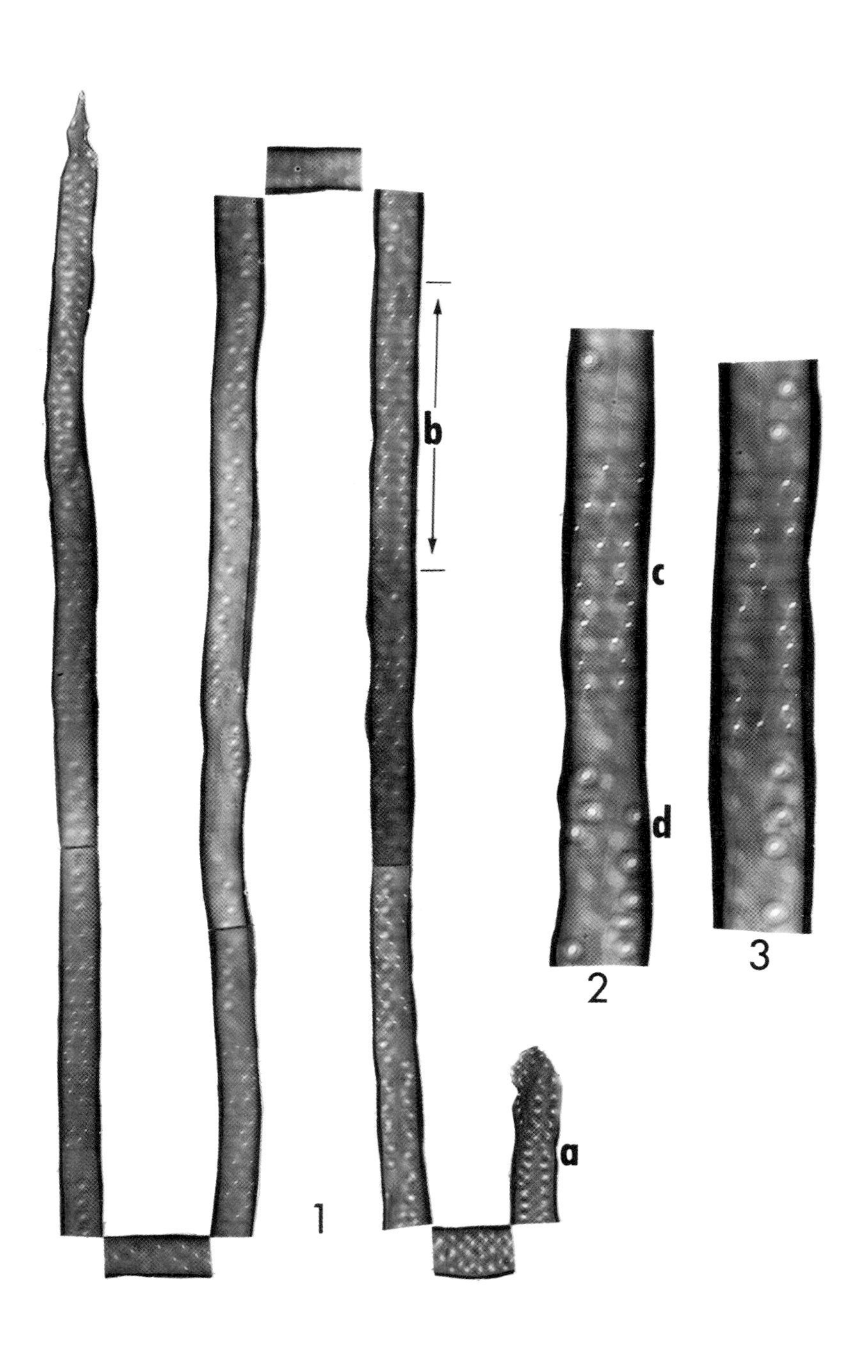
b
c
d
a
1
2
3

REDWOOD

Sequoia sempervirens (D. Don) Endl.

1. Springwood longitudinal tracheid, 95X, radial view
   - *a.* Intertracheid pits: bordered pits of pit pairs leading to a contiguous longitudinal tracheid
   - *b.* Ray-contact area
2. Portion of a longitudinal tracheid, 185X
   - *c.* Ray-parenchyma pits: small, elliptical, uniformly sized pits of pit pairs leading to ray parenchyma
   - *d.* Intertracheid pits: bordered pits of pit pairs leading to a contiguous longitudinal tracheid

PLATE 20

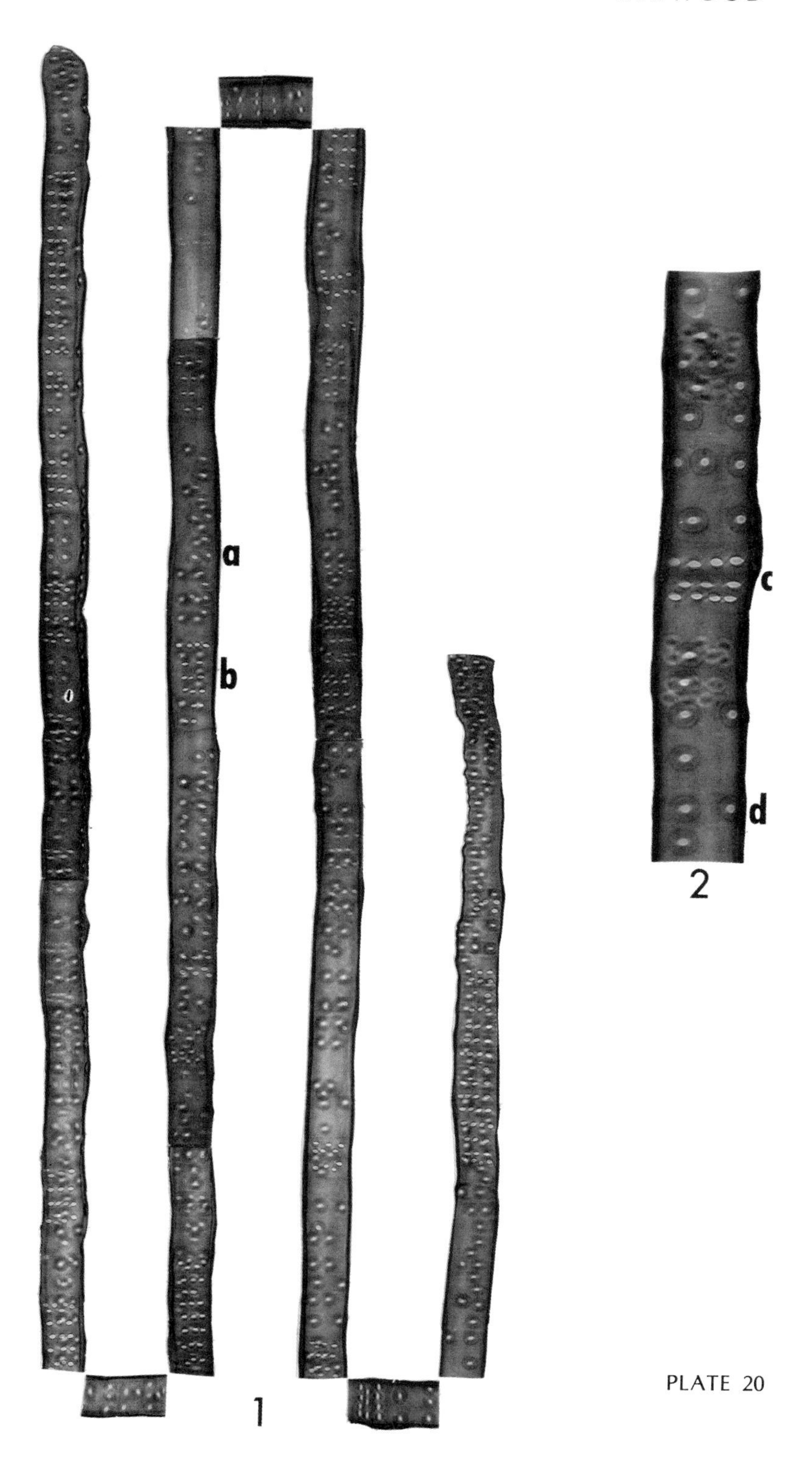

PLATE 20

BALDCYPRESS

Taxodium distichum (L.) Rich.

1. Springwood longitudinal tracheid, 95X, radial view
   - *a.* Intertracheid pits: bordered pits of pit pairs leading to a contiguous longitudinal tracheid
   - *b.* Ray-contact area
2. Portion of a longitudinal tracheid, 185X
   - *c.* Ray-parenchyma pits: small, elliptical, uniformly sized pits of pit pairs leading to ray parenchyma
   - *d.* Intertracheid pits: bordered pits of pit pairs leading to a contiguous longitudinal tracheid

INCENSE-CEDAR

Libocedrus decurrens Torr.
[*Calocedrus decurrens* (Torr.) Florin]

1. Springwood longitudinal tracheid, 95X, radial view
   - *a.* Intertracheid pits: bordered pits of pit pairs leading to a contiguous longitudinal tracheid
   - *b.* Ray-contact area
2. Portion of a longitudinal tracheid, 185X
   - *c.* Ray-parenchyma pits: small, elliptical, uniformly sized pits of pit pairs leading to ray parenchyma
   - *d.* Intertracheid pits: bordered pits of pit pairs leading to a contiguous longitudinal tracheid

PLATE 21

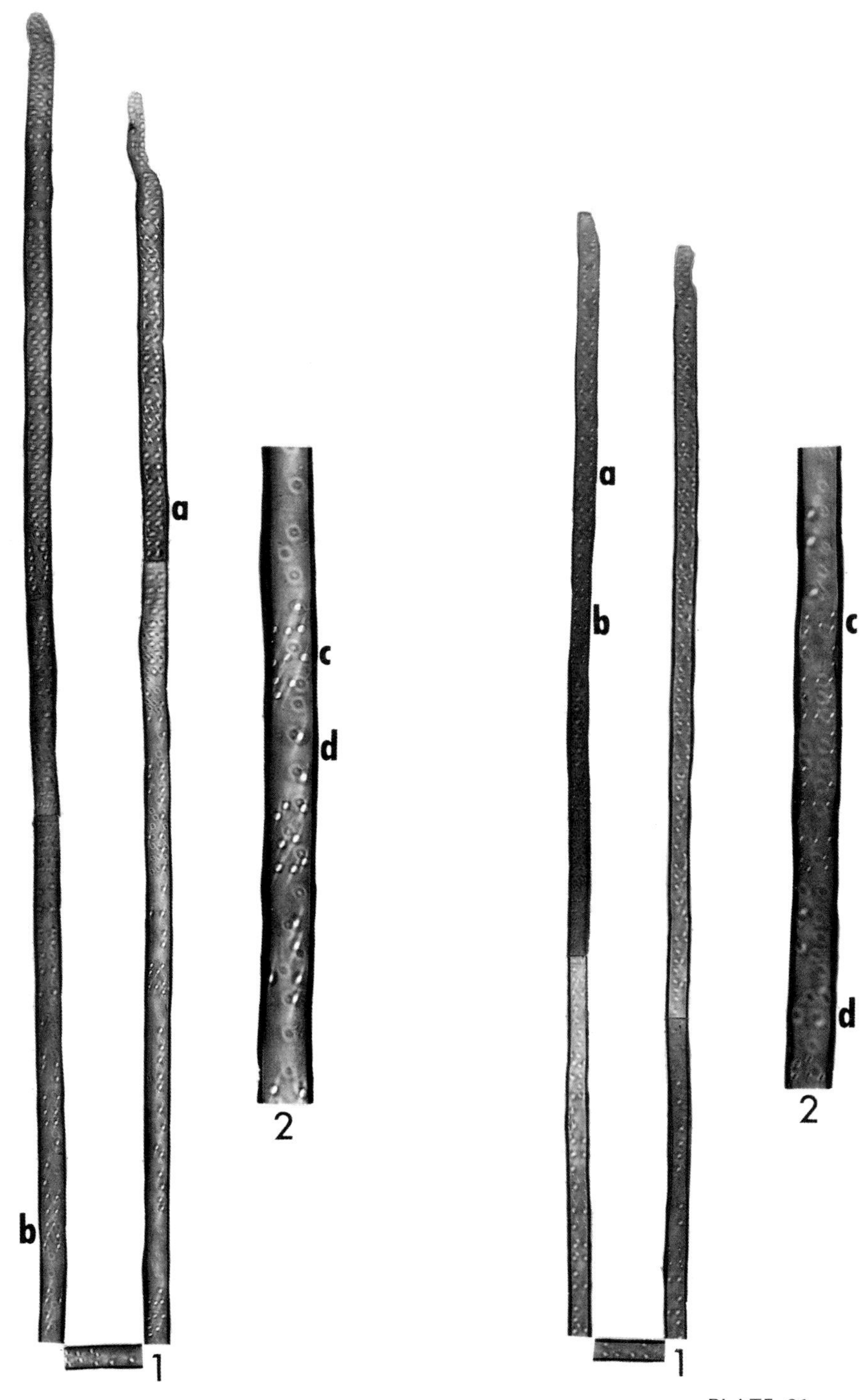
BALDCYPRESS
INCENSE-CEDAR
a
b
c
d
1
2
a
b
c
d
1
2

WESTERN REDCEDAR
Thuja plicata Donn

1. Springwood longitudinal tracheid, 95X, radial view
   - *a.* Intertracheid pits: bordered pits of pit pairs leading to a contiguous longitudinal tracheid
   - *b.* Ray-contact area
2. Portion of a longitudinal tracheid, 185X
   - *c.* Ray-parenchyma pits: small, elliptical, uniformly sized pits of pit pairs leading to ray parenchyma
   - *d.* Intertracheid pits: bordered pits of pit pairs leading to a contiguous longitudinal tracheid
3. Portion of a longitudinal tracheid, 185X, showing pit variation

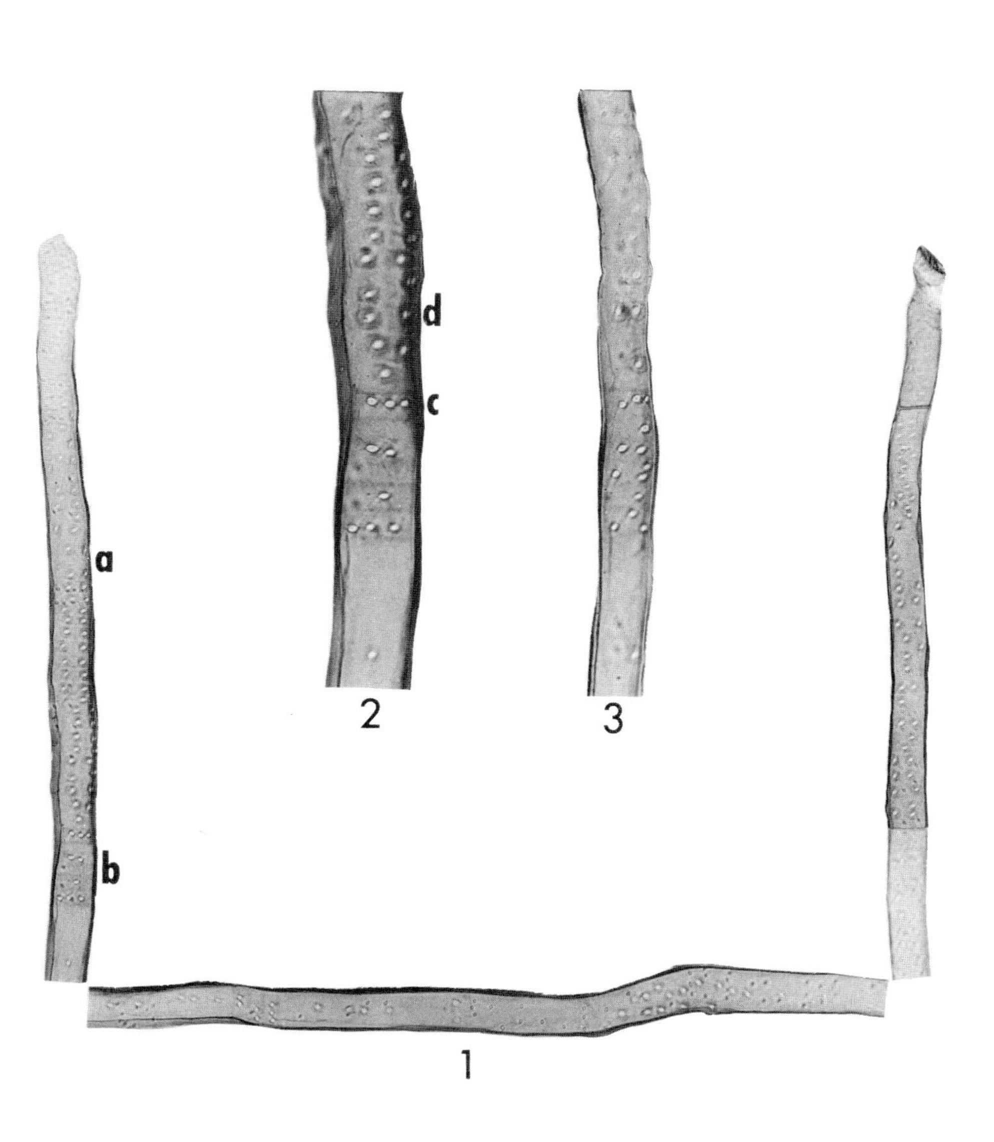
a
b
d
c
2
3
1

PORT-ORFORD-CEDAR

Chamaecyparis lawsoniana (A. Murr.) Parl.

1. Springwood longitudinal tracheid, 95X, radial view
   a. Intertracheid pits: bordered pits of pit pairs leading to a contiguous longitudinal tracheid
   b. Ray-contact area

2. Portion of a longitudinal tracheid, 185X
   c. Ray-parenchyma pits: small, elliptical, uniformly sized pits of pit pairs leading to ray parenchyma
   d. Intertracheid pits: bordered pits of pit pairs leading to a contiguous longitudinal tracheid

3. Portion of a longitudinal tracheid, 185X

ALASKA-CEDAR

Chamaecyparis nootkatensis (D. Don) Spach

1. Springwood longitudinal tracheid, 95X, radial view
   a. Intertracheid pits: bordered pits of pit pairs leading to a contiguous longitudinal tracheid
   b. Ray-contact area

2. Portion of a longitudinal tracheid, 185X
   c. Ray-parenchyma pits: small, elliptical, uniformly sized pits of pit pairs leading to ray parenchyma
   d. Intertracheid pits: bordered pits of pit pairs leading to a contiguous longitudinal tracheid

3. Portion of a longitudinal tracheid, 185X, showing pit variation

PORT-ORFORD-CEDAR | ALASKA-CEDAR

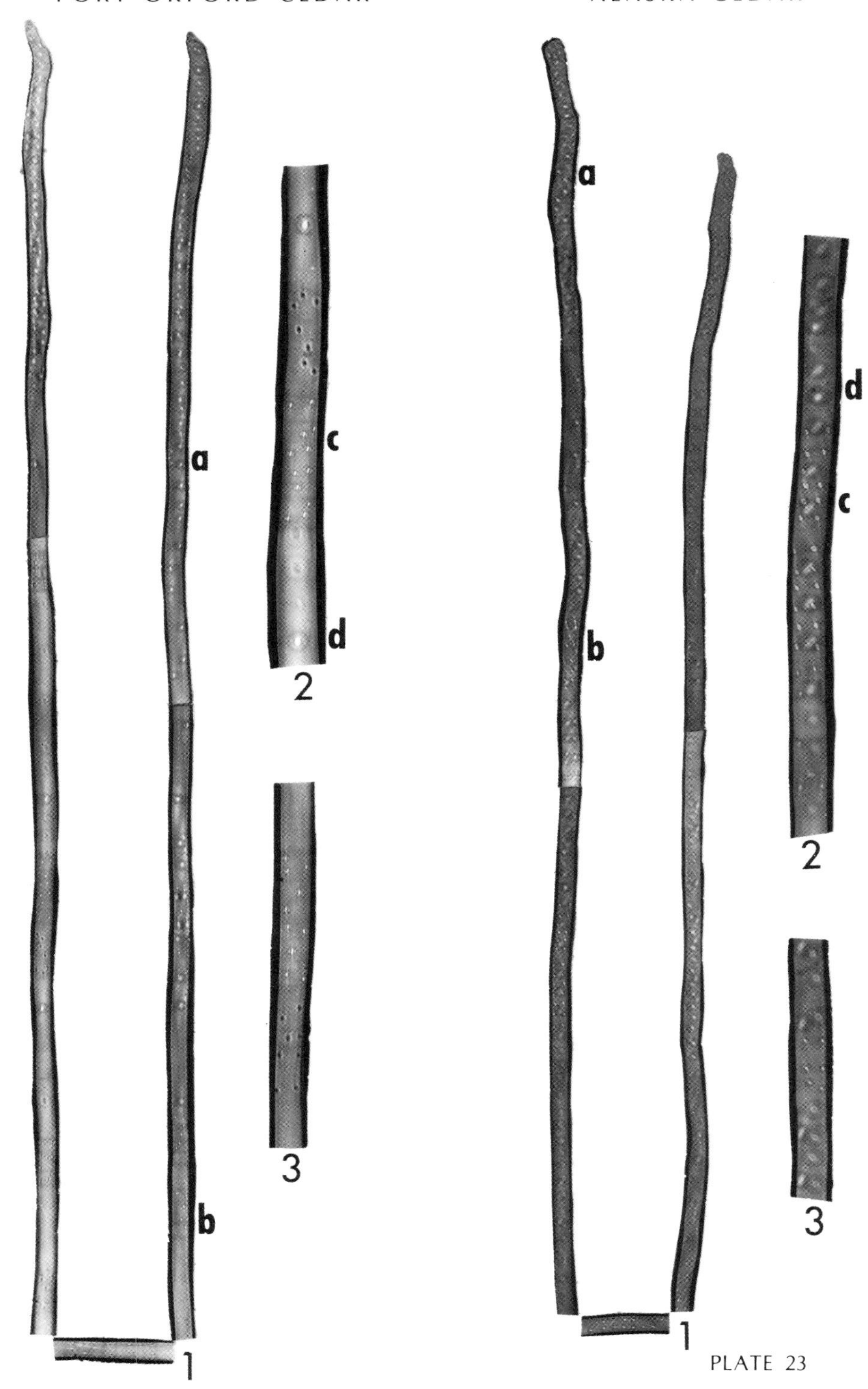

PLATE 23

PARANA-PINE

Araucaria angustifolia (Bert.) O. Ktze.

1. Springwood longitudinal tracheid, 95X, radial view
   - *a.* Intertracheid pits: bordered pits of pit pairs leading to a contiguous longitudinal tracheid
   - *b.* Ray-contact area
2. Portion of a longitudinal tracheid, 185X
   - *c.* Ray-parenchyma pits: large, varishaped (often slit-like) pits of pit pairs leading to ray parenchyma
3. Portion of a longitudinal tracheid, 185X, showing pit variation

PLATE 24

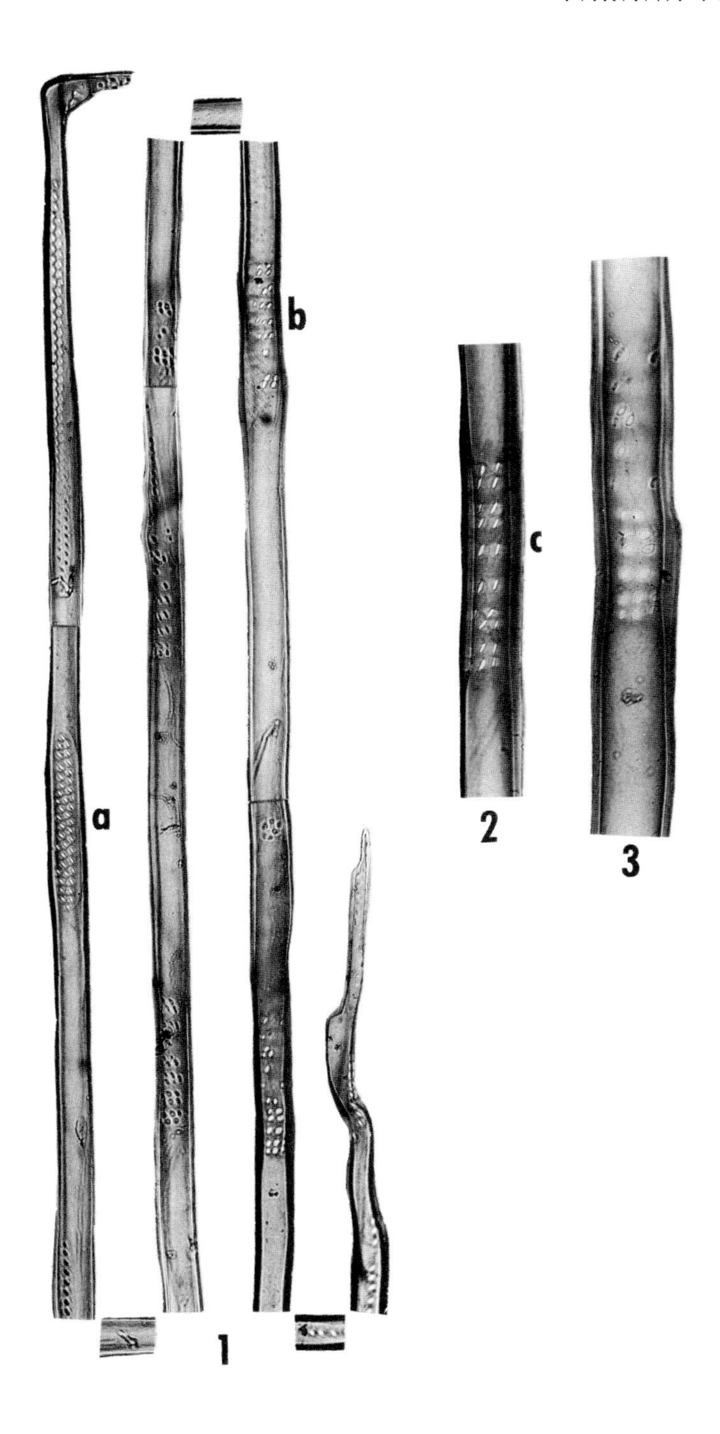
a
b
c
1
2
3

BLACK WILLOW

Salix nigra Marsh.

1. Vessel element, 185X
   - *a.* Ray-contact area consisting of three longitudinally elongated ray-cell contact areas
   - *b.* Simple perforation
2. Vessel element, 185X
   - *c.* Intervessel pitting
3. Vessel element, 185X
   - *d.* Simple perforation
   - *e.* Intervessel pitting
4. Vessel element, 185X, showing intervessel pitting (caudate tip) and three ray-contact areas
5. Vessel element, 185X
   - *f.* Ray-contact area: four areas are visible at different levels
6. Fiber, 185X

PLATE 25

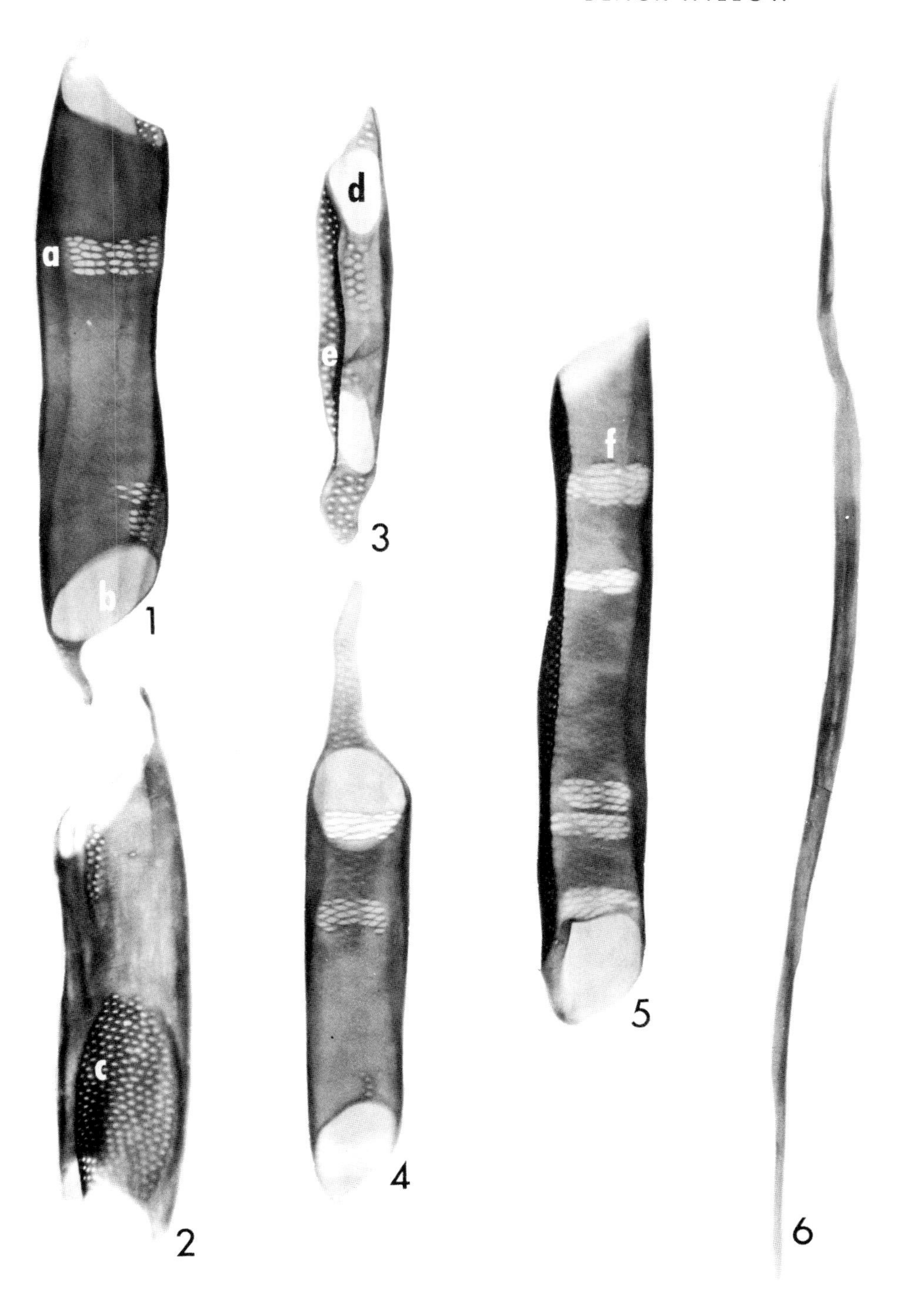

PLATE 25

BIGTOOTH ASPEN

Populus grandidentata Michx.

1. Vessel element, 185X
   - *a.* Simple perforation
   - *b.* Ray-contact area: two such areas are visible
2. Vessel element, 185X
   - *c.* Simple perforation
3. Vessel element, 185X
   - *d.* Intervessel pitting
4. Vessel element, 185X
   - *e.* Intervessel pitting
5. Fiber, 185X

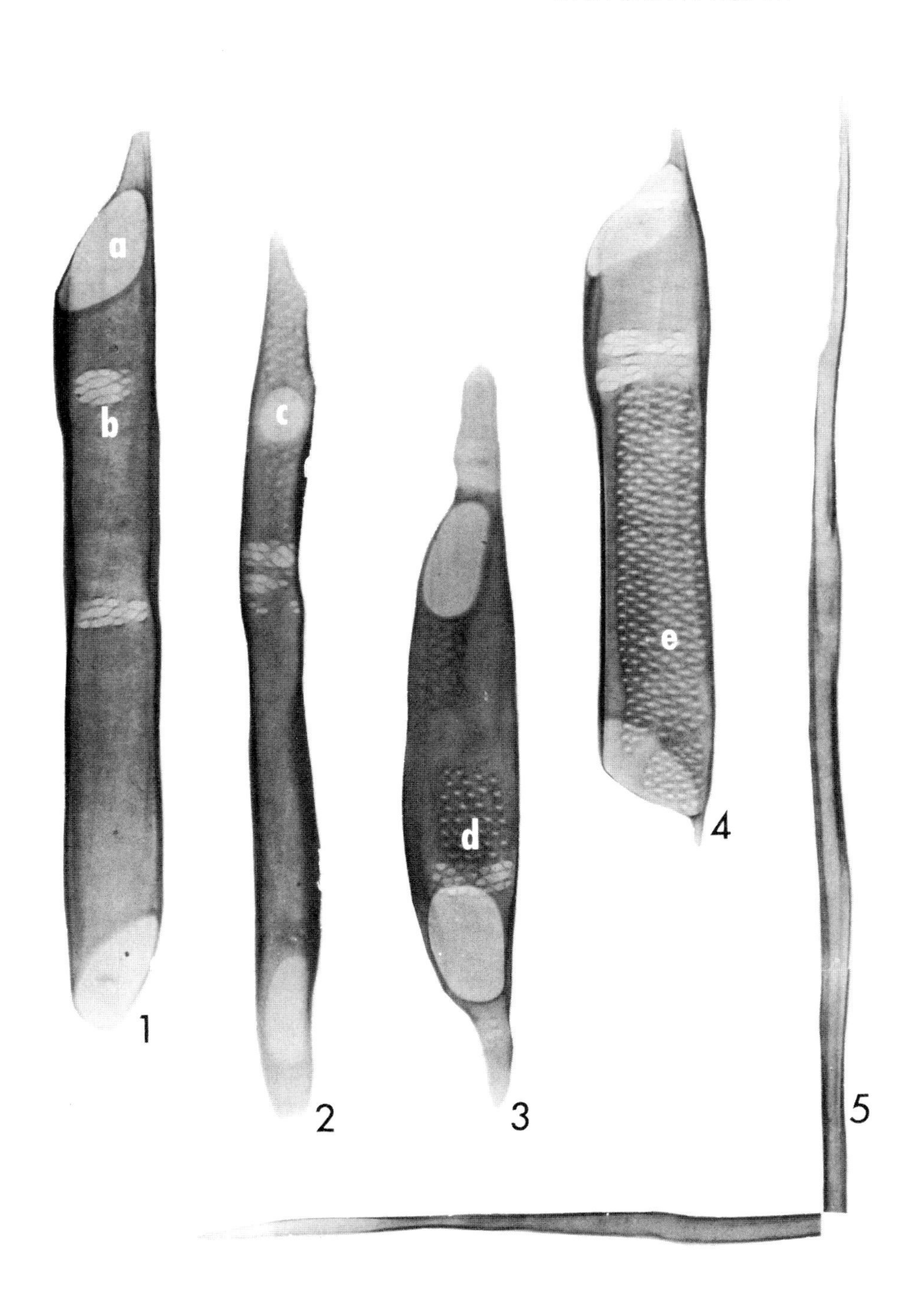

PLATE 26

EUROPEAN POPLAR
Populus tremula L.

1. Vessel element, 95X
    a. Intervessel pitting
2. Vessel element, 95X
    b. Simple perforation
3. Vessel element, 95X
    c. Ray contact area
4. Vessel element, 95X, with tyloses
5. Vessel element, 95X, with tyloses
6. Vessel element, 95X
    d. Intervessel pitting
7. Fiber, 95X

PLATE 27

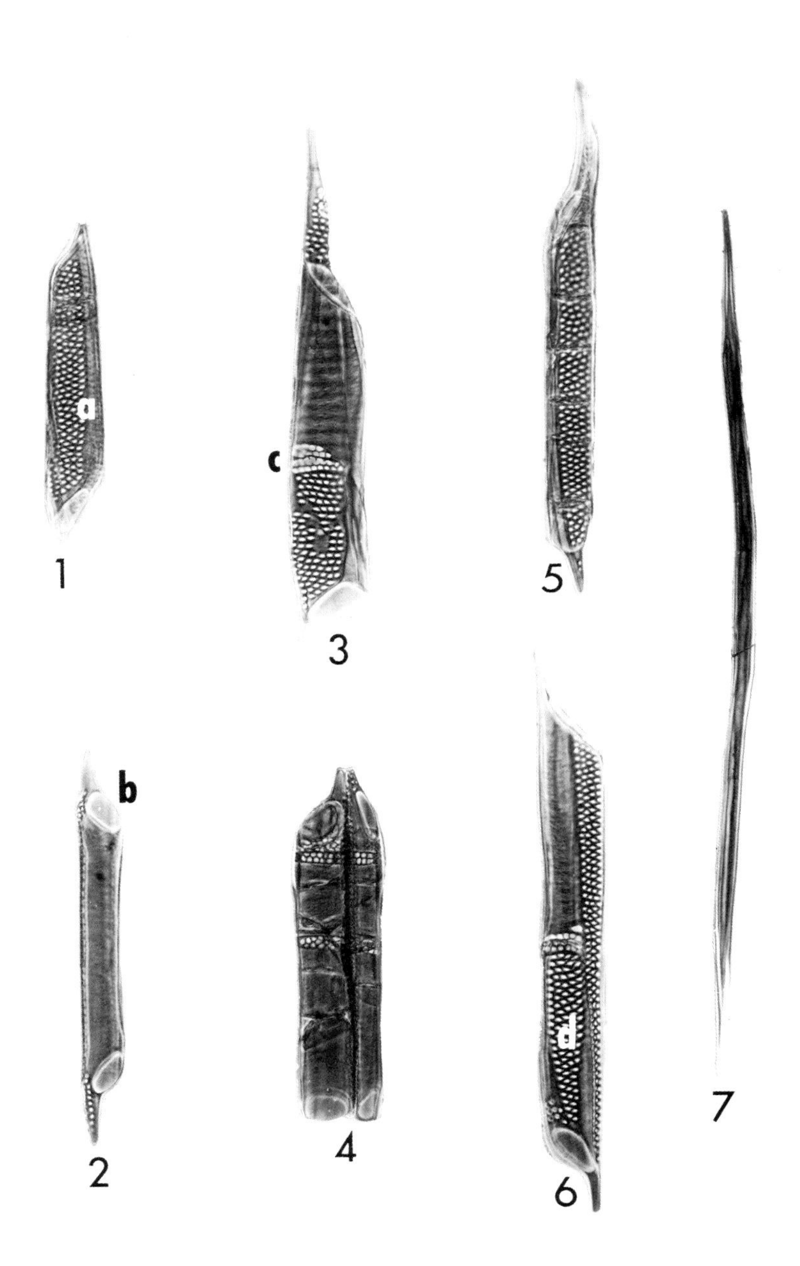
a
c
b
d
1
3
5
2
4
6
7

BALSAM POPLAR

Populus balsamifera L.

1. Vessel element, 95X
   - *a.* Ray-contact area
   - *b.* Simple perforation
2. Vessel element, 95X
   - *c.* Area of contact with longitudinal parenchyma
3. Vessel element, 95X
   - *d.* Intervessel pitting
4. Vessel element, 95X
5. Vessel element, 95X
6. Vessel element, 95X
7. Fiber, 95X

PLATE 28

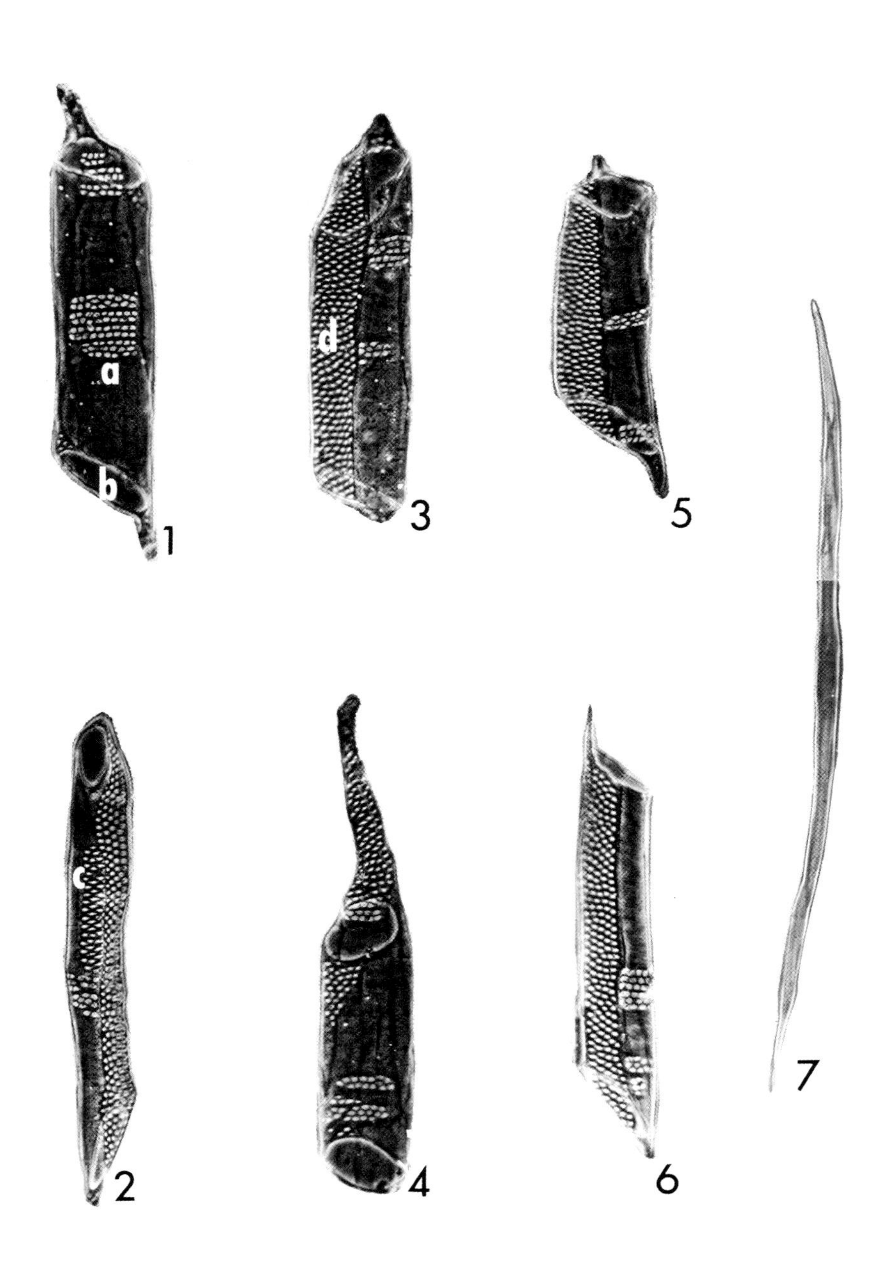
a
b
1
c
2
d
3
4
5
6
7

EASTERN COTTONWOOD

Populus deltoides Bartr.

1. Vessel element, 185X
    - *a.* Ray-contact area: three such areas are visible
    - *b.* Simple perforation
2. Vessel element, 185X
    - *c.* Area of contact with longitudinal parenchyma
3. Vessel element, 185X
    - *d.* Intervessel pitting
4. Vessel element, 185X
5. Fiber, 185X

PLATE 29

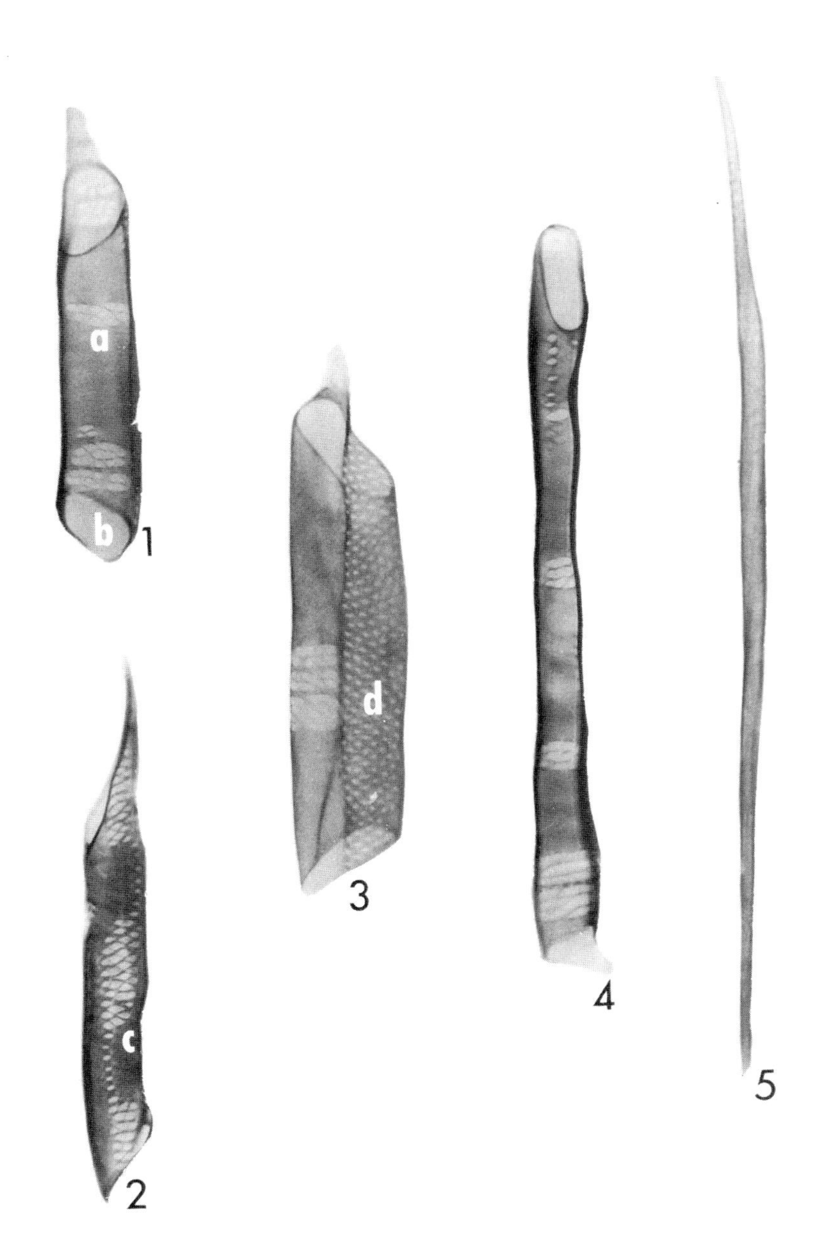
a
b
1
2
c
3
d
4
5

BLACK COTTONWOOD

Populus trichocarpa Torr. and Gray

1. Vessel element, 95X
   - *a.* Intervessel pitting
   - *b.* Simple perforation
2. Vessel element, 95X
   - *c.* Ray-contact area
   - *d.* Simple perforation: intervessel pits visible in background
3. Vessel element, 95X
   - e. Ray-contact area
4. Vessel element, 95X
   - *f.* Intervessel pitting
5. Fiber, 95X
6. Fiber, 95X

PLATE 30

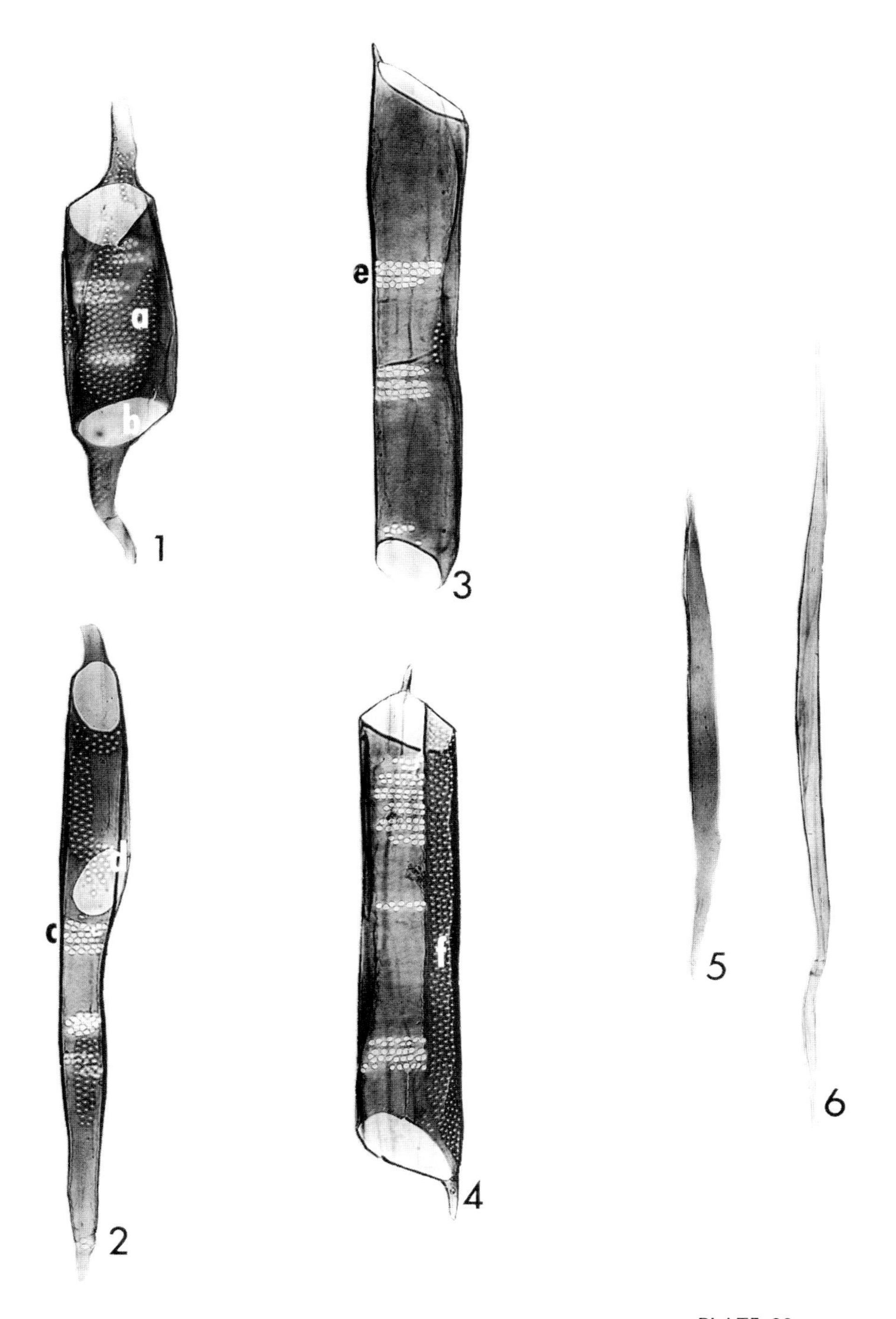

PLATE 30

BLACK WALNUT
Juglans nigra L.

1. Summerwood vessel element, 95X
   a. Intervessel pitting: vessel element is reticulate on the back wall
2. Vessel element, 95X: element is reticulate on the front wall
3. Springwood vessel element, 95X: two tyloses are visible
   b. Ray-contact area in the foreground
4. Springwood vessel element, 95X
   c. Cells of longitudinal parenchyma in contact, as in the wood
5. Fiber, 95X

PLATE 31

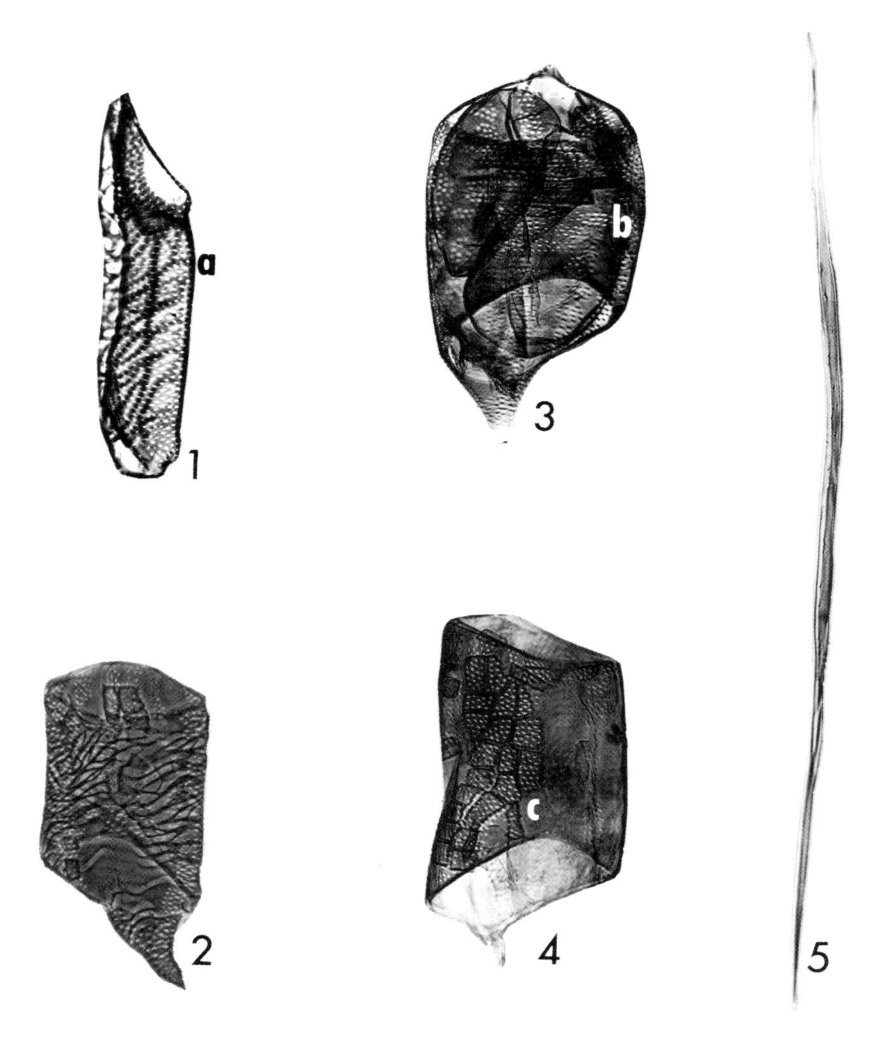
a
1
b
3
2
c
4
5

BUTTERNUT
Juglans cinerea L.

1. Vessel element, 95X
   a. Ray-contact area
2. Vessel element, 185X
   b. Simple perforation
   c. Intervessel pitting
   d. Area of contact with longitudinal parenchyma
   e. Ray-contact area
3. Vessel element, 95X, tangential view
   f. Intervessel pitting
4. Vessel element, 95X, showing areas of contact with longitudinal parenchyma
5. Fiber, 95X

PLATE 32

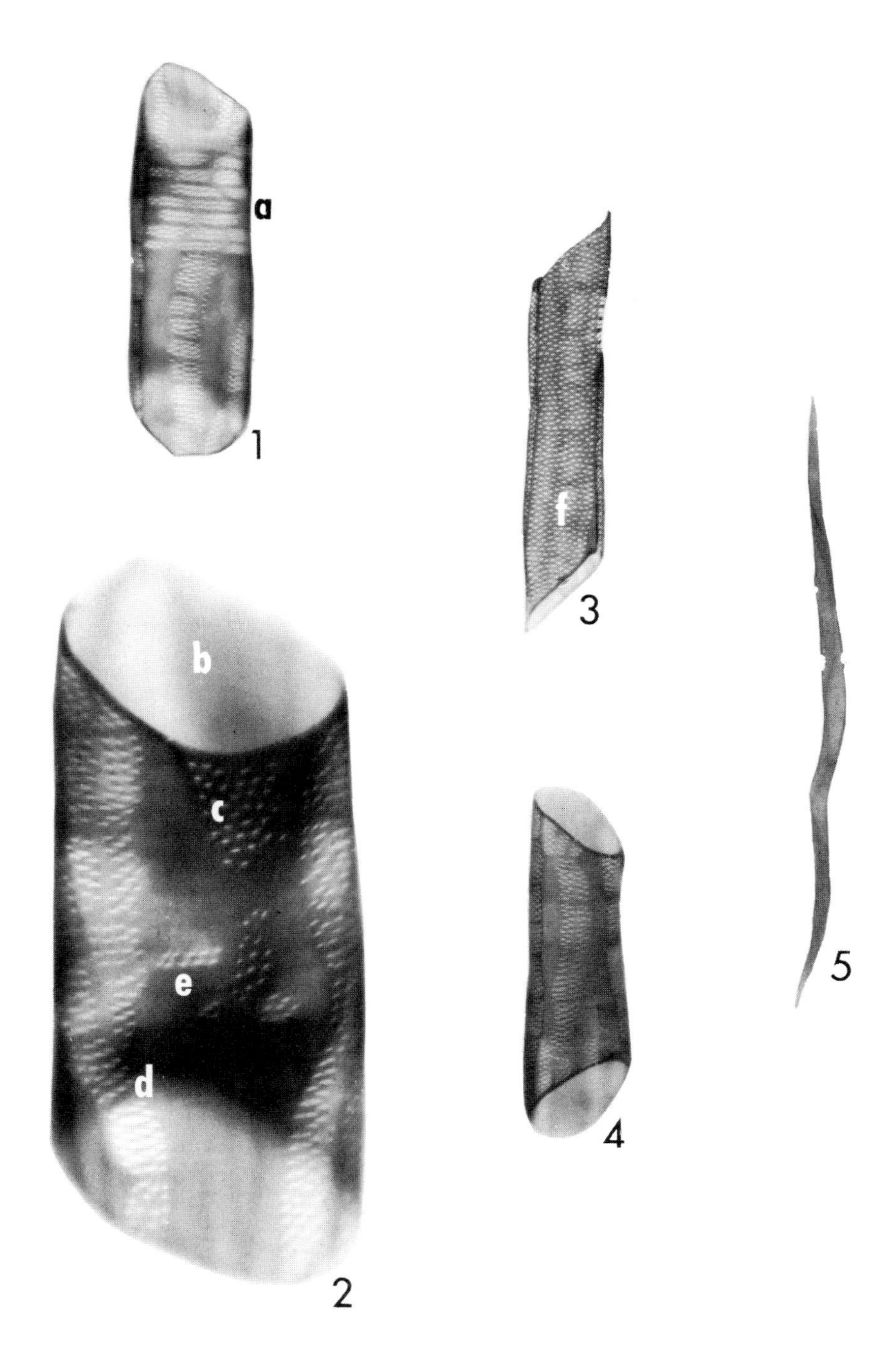

PLATE 32

SHAGBARK HICKORY

Carya ovata (Mill.) K. Koch

1. Summerwood vessel element, 95X, with simple perforation at each end
2. Summerwood vessel element, 95X
   - *a.* Simple perforation
3. Late springwood vessel elements, 95X, showing intervessel pitting
4. Six springwood vessel elements, 95X, in contact, as in the wood
   - *b.* Longitudinal parenchyma in contact, as in the wood
5. Two vessel elements, 95X, partially separated, containing tyloses
   - *c.* Area of contact with longitudinal parenchyma
   - *d.* Ray-contact area
   - *e.* Displaced ray cells
6. Fiber, 95X

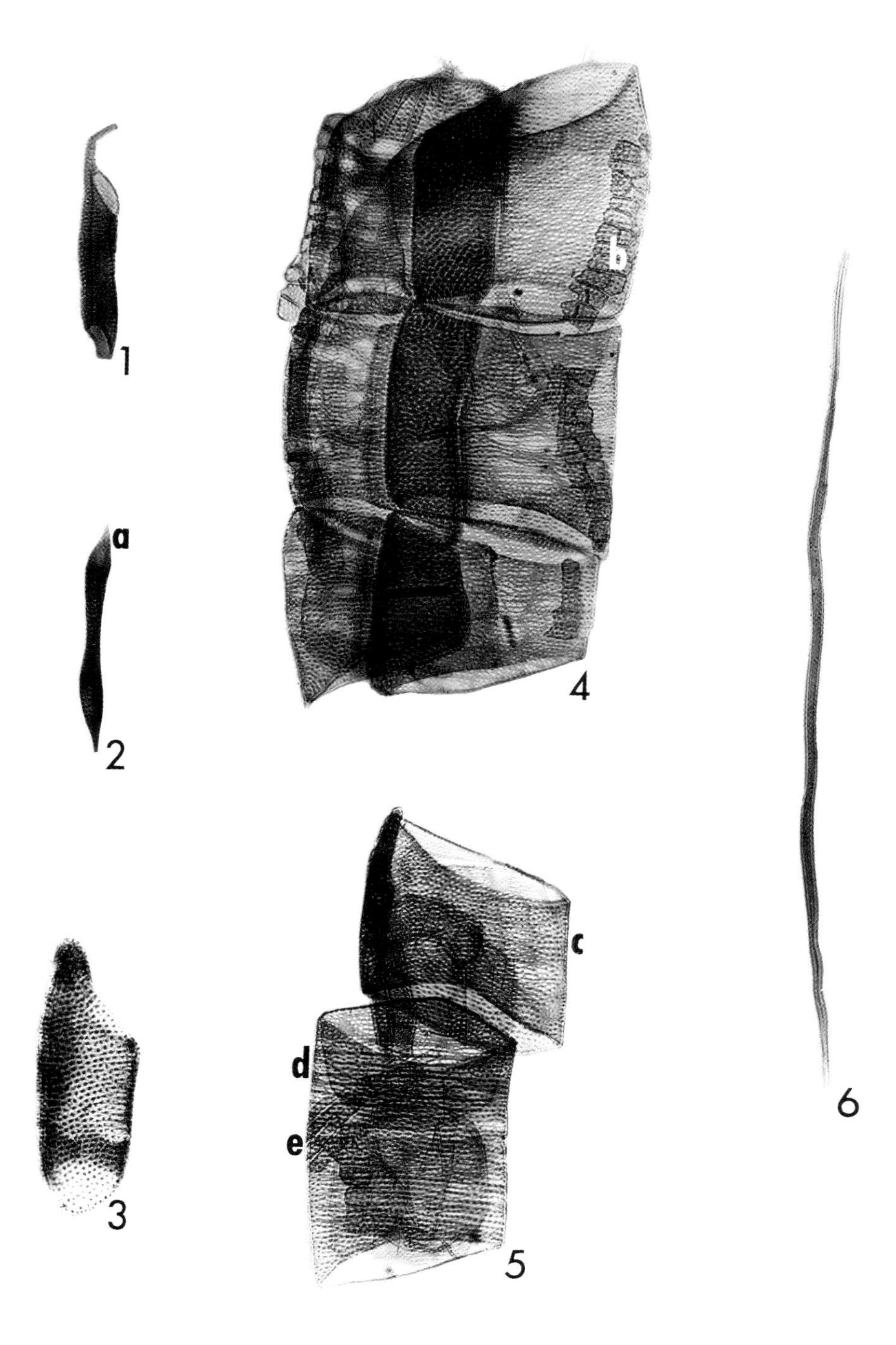

PLATE 33

YELLOW BIRCH

Betula alleghaniensis Britton

1. Vessel element, 185X
   - *a.* Intervessel pitting: pits are extremely fine and confluent
   - *b.* Ray-contact area: pits are extremely fine and confluent
   - *c.* Area of contact with longitudinal parenchyma: pits are extremely fine and confluent
   - *d.* Scalariform perforation with many thin bars
2. Vessel element, 95X
3. Narrow vessel element, 95X
4. Fiber, 95X

PLATE 34

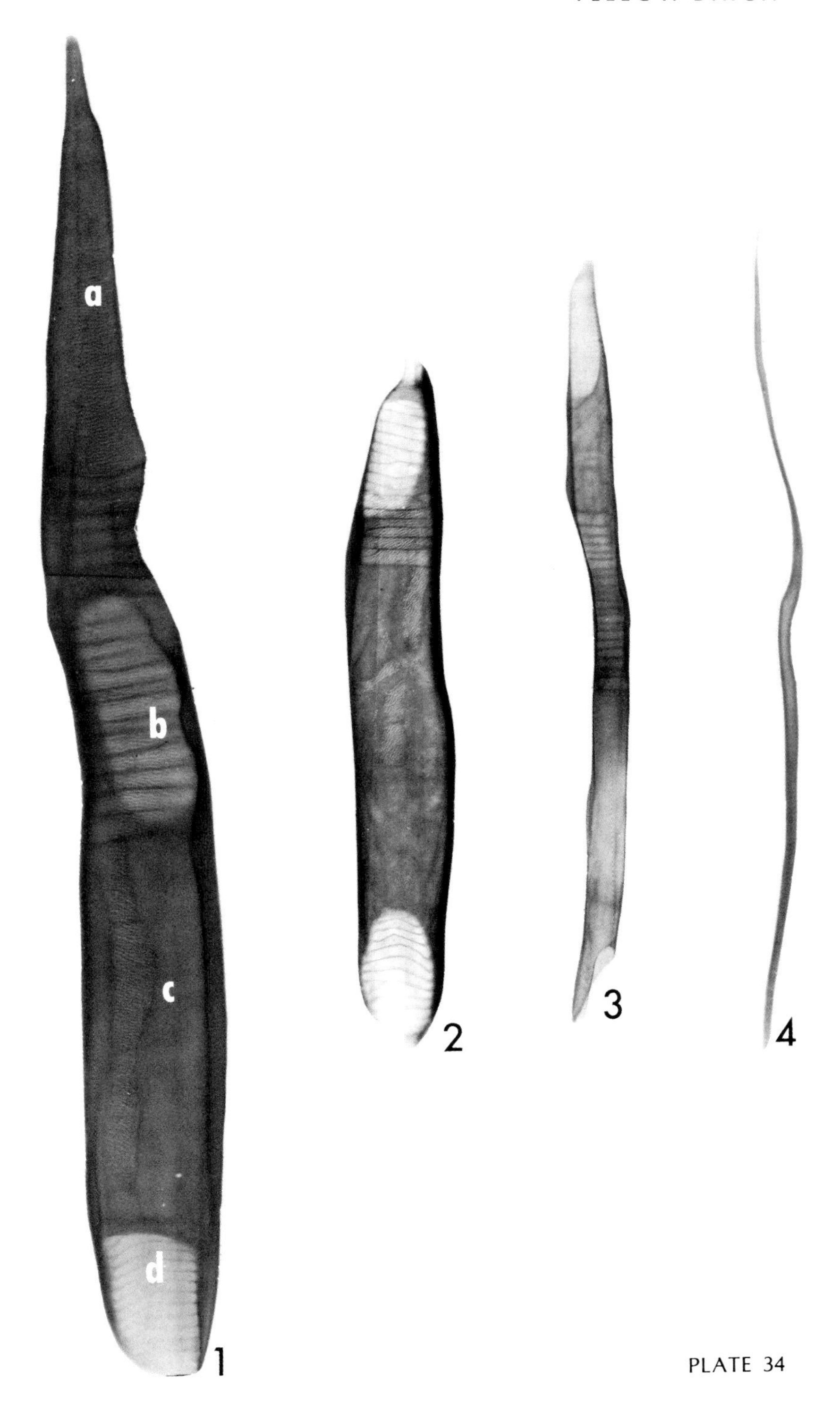

PLATE 34

PAPER BIRCH, WHITE BIRCH

Betula papyrifera Marsh.

1. Vessel element, 95X
   - *a.* Scalariform perforation with many thin bars
   - *b.* Ray-contact area
2. Vessel element, 95X
   - *c.* Scalariform perforation with many thin bars
   - *d.* Ray-contact area: pits are extremely fine
   - e. Area of contact with longitudinal parenchyma
3. Three vessel elements, 95X, in contact, as in the wood
   - *f.* Intervessel pitting: pits are extremely fine
4. Vessel element, 95X
   - *g.* Intervessel pitting: pits are extremely fine
5. Fiber, 95X

PLATE 35

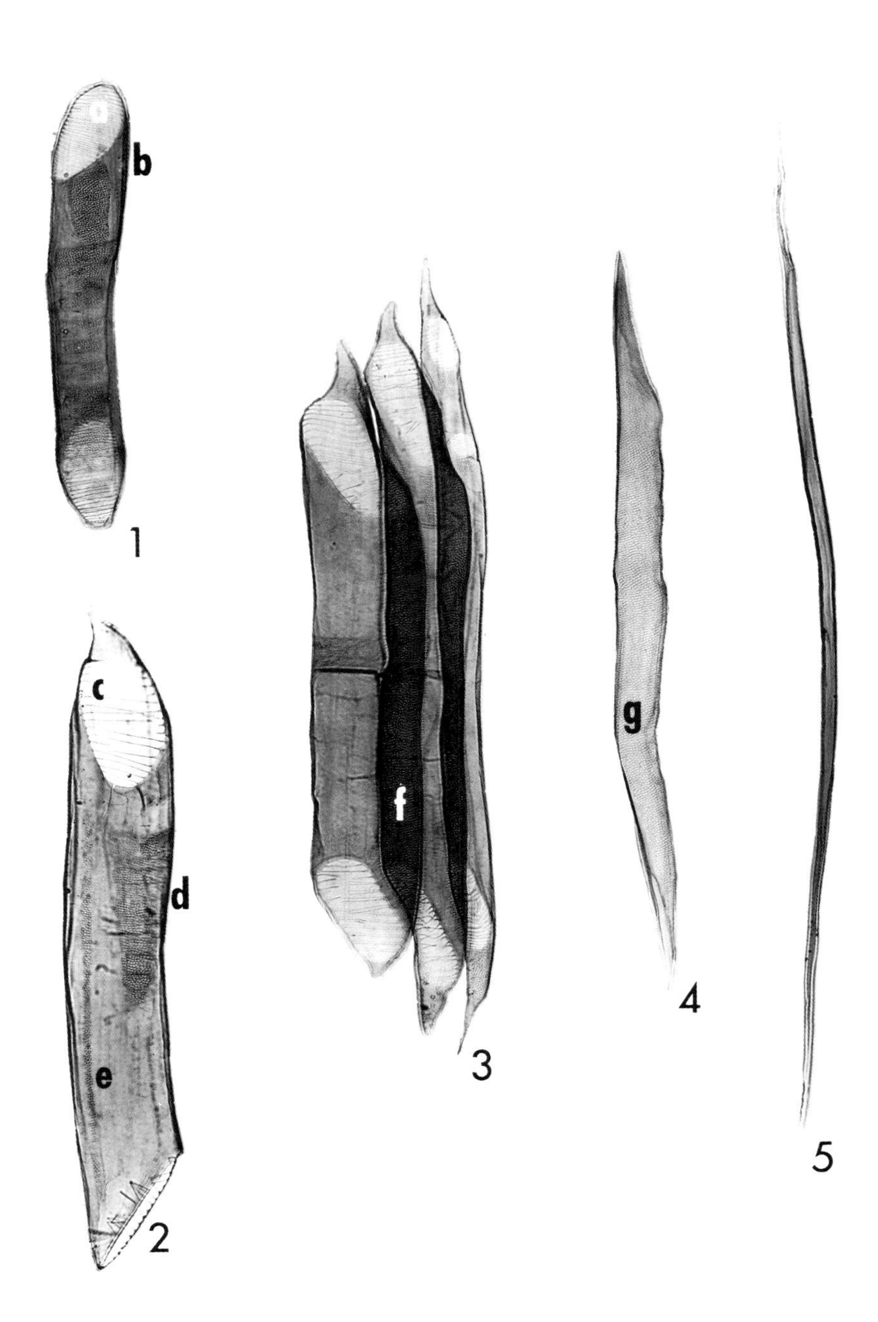

PLATE 35

EUROPEAN BIRCH

Betula verrucosa Ehrh.

1. Vessel element, 95X
   - *a.* Intervessel pitting: Pits are extremely fine and confluent
2. Vessel element, 95X
3. Vessel elements, 95X
   - *b.* Ray-contact area: Pits are extremely fine and confluent
   - *c.* Scalariform perforation with many thin parts
4. Vessel element, 95X
   - *d.* Area of contact with longitudinal parenchyma
5. Fiber, 95X
6. Fiber, 95X

PLATE 36

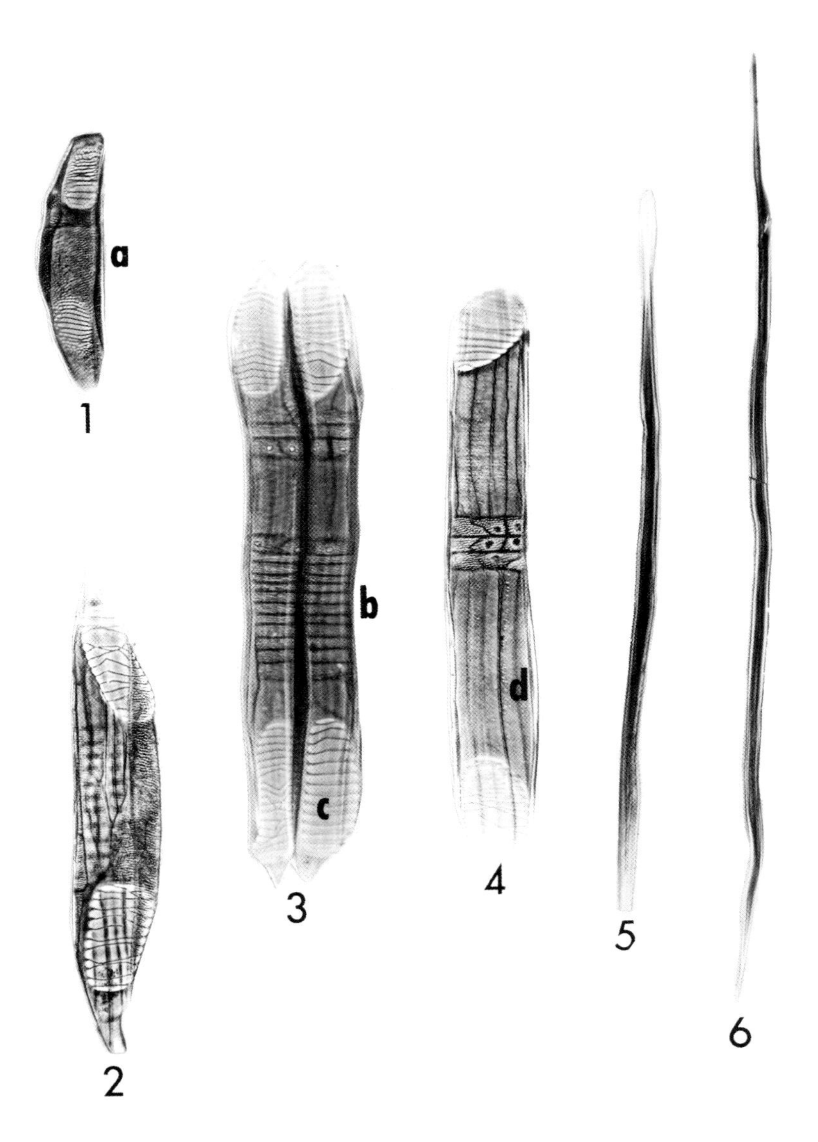

PLATE 36

RED ALDER
Alnus rubra Bong.

1. Vessel element, 95X
   - *a.* Scalariform perforation with many thin bars
   - *b.* Ray-contact area
   - *c.* Area of contact with longitudinal parenchyma
2. Vessel element, 95X
   - *d.* Ray-contact area
3. Vessel element, 95X, with three scalariform perforations
4. Vessel element, 95X, with scalariform perforations, ray-contact areas, and area of contact with longitudinal parenchyma
5. Vessel element, 95X
   - *e.* Intervessel pitting
6. Fiber, 95X

PLATE 37

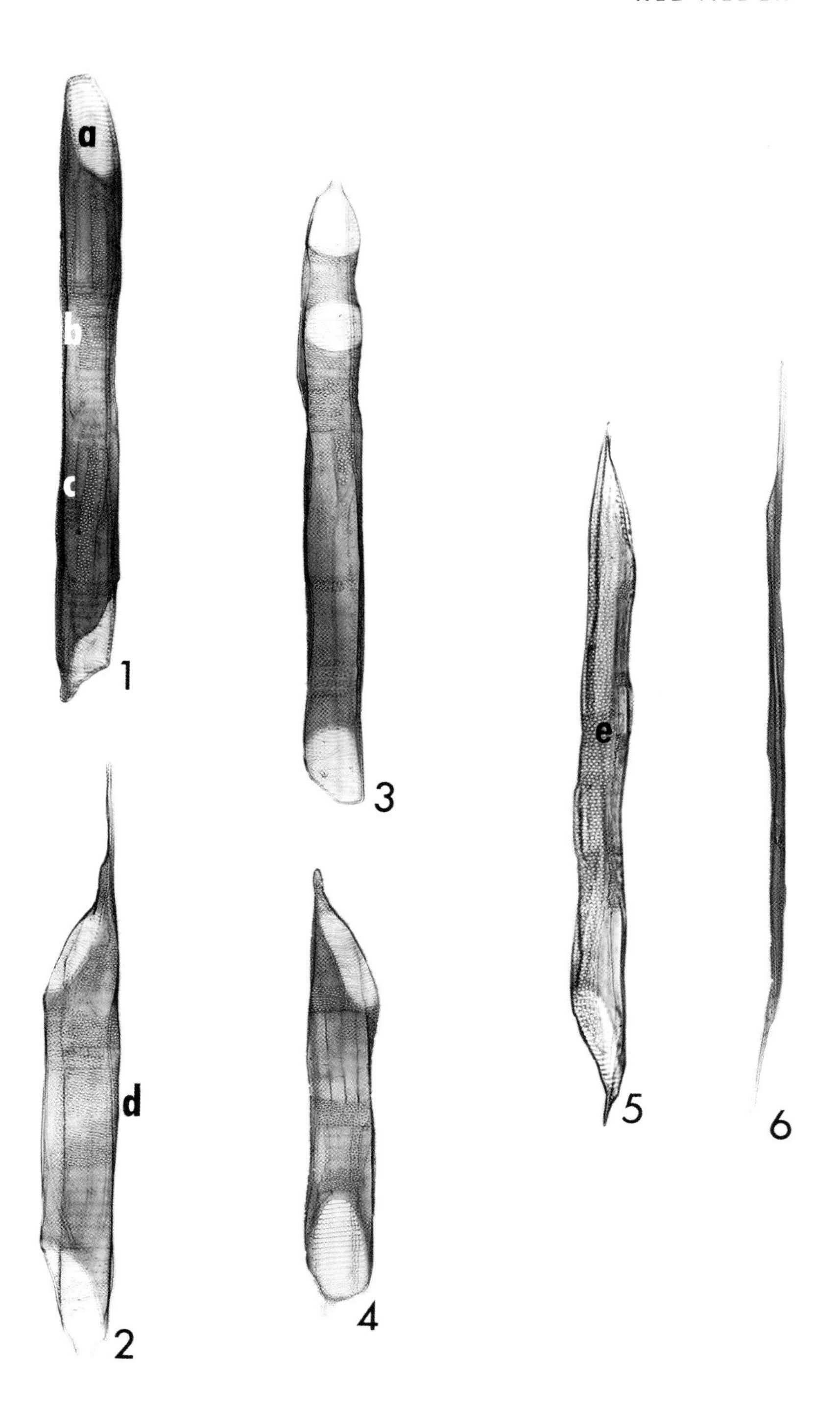
a
b
c
1
2
d
3
4
e
5
6

BEECH

Fagus grandifolia Ehrh.

1. Vessel element, 185X
   - *a.* Simple perforation
   - *b.* Ray-contact area
2. Vessel elements, 185X
   - *c.* Fiber-contact area
   - *d.* Intervessel pitting
3. Portion of vessel element, 185X
   - e. Scalariform perforation
4. Fiber, 185X
5. Fiber, 185X

PLATE 38

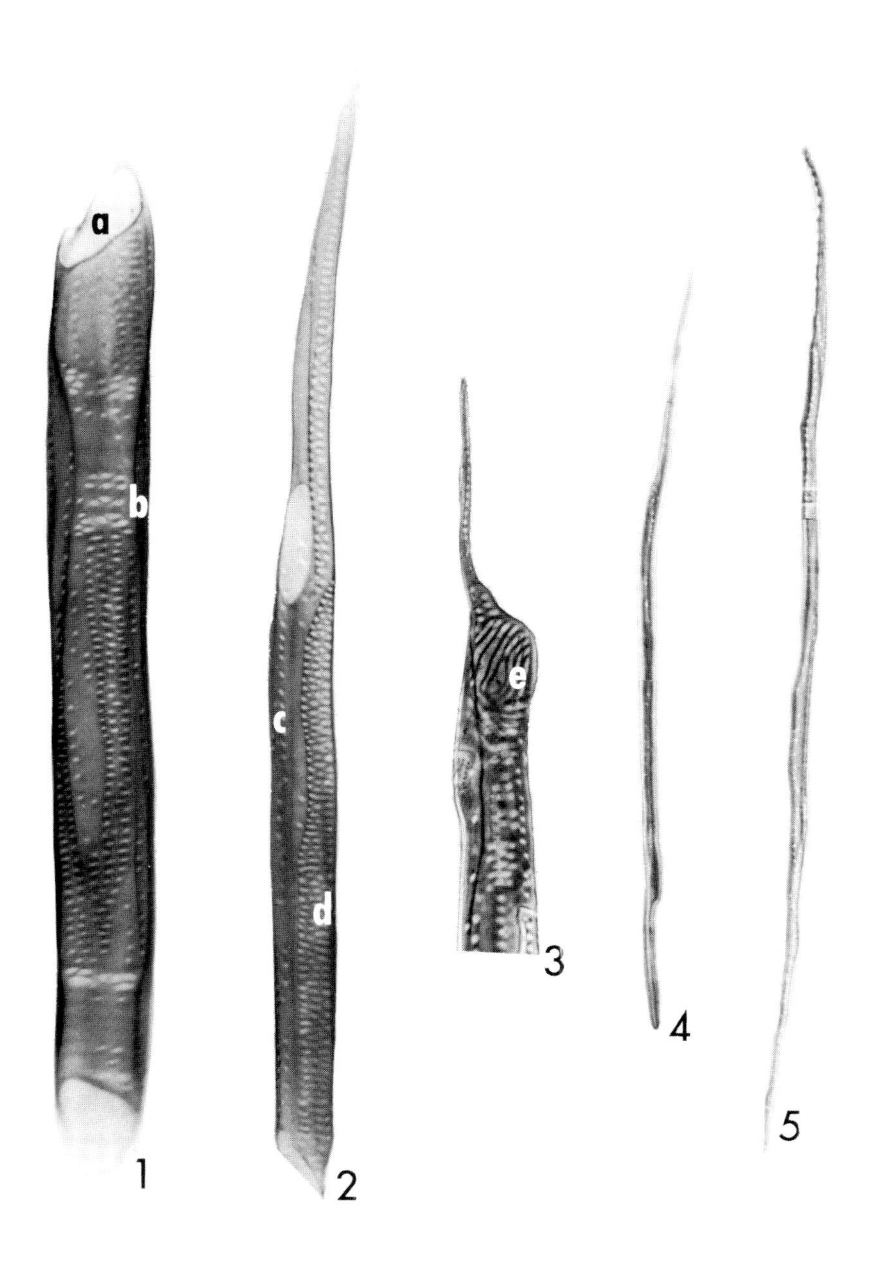
a
b
c
d
e
1
2
3
4
5

CHESTNUT
Castanea dentata (Marsh.) Borkh.

1. Springwood vessel element, 95X
   a. Ray-contact area
2. Springwood vessel element, 95X
3. Springwood vessel element, 185X
   b. Simple perforation
   c. Ray-contact area
   d. Areas of contact with vasicentric tracheids
4. Summerwood vessel element, 95X
   e. Scalariform perforation (rare) with one bar
5. Vasicentric tracheid, 95X
6. Vasicentric tracheid, 95X
7. Fiber, 95X

PLATE 39

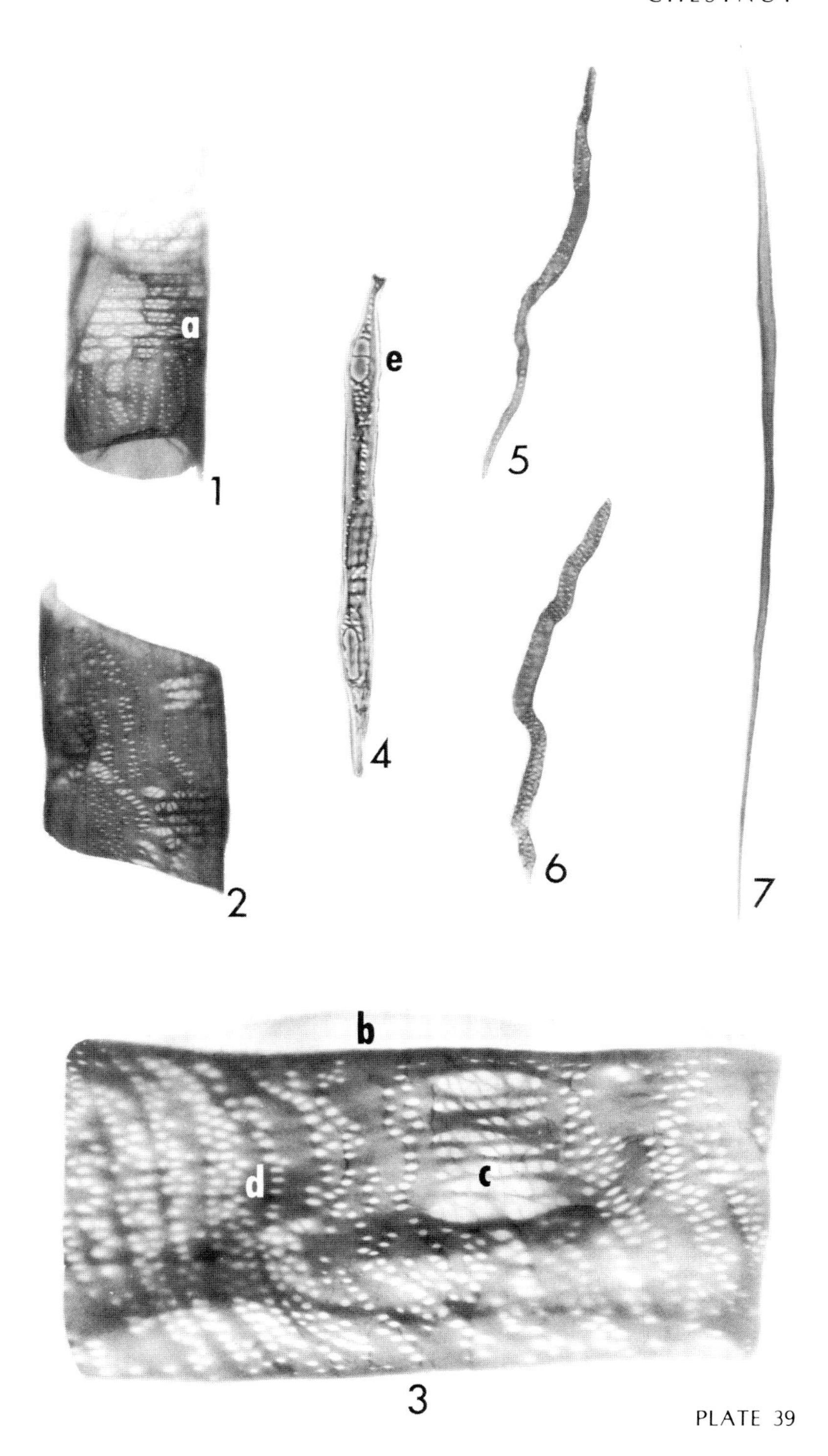
a
e
1
2
3
4
5
6
7
b
c
d

WHITE OAK
Quercus alba L.

1. Vasicentric tracheid, 95X
    a. Ray-contact area
2. Summerwood vessel element, 95X
3. Vessel element, 95X, containing a balloon-like tylosis
    b. Ray-contact area
4. Summerwood vessel element, 95X
    c. Simple perforation
5. Springwood vessel element, 95X, containing tyloses
6. Portion of large springwood vessel element, 95X
    d. Ray-contact area
    e. Intervessel pitting
    f. Area in contact with vasicentric tracheids
7. Summerwood vessel element, 95X, with three simple perforations
8. Fiber, 95X

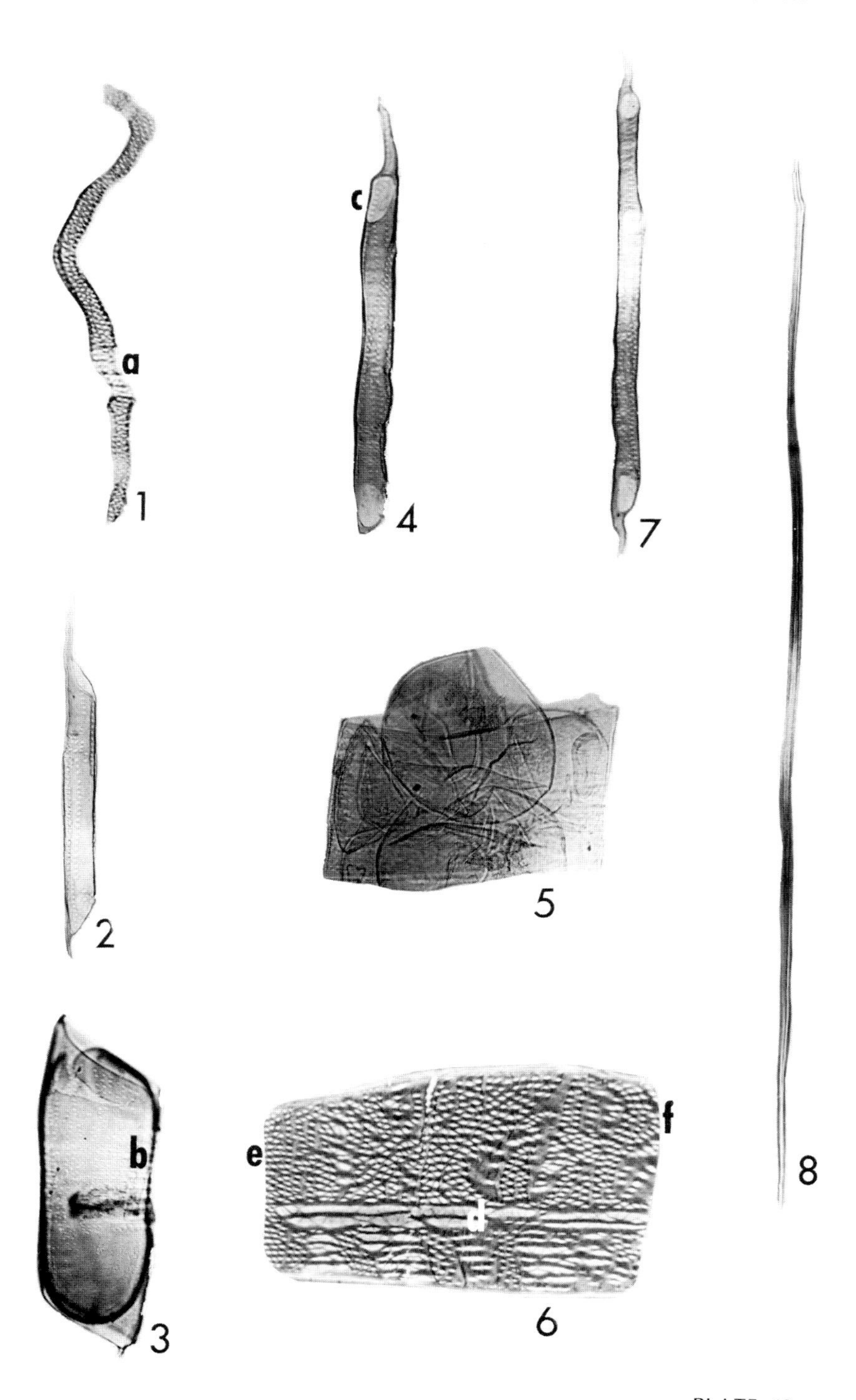
a
1
c
4
7
2
5
b
3
e
f
d
6
8

NORTHERN RED OAK

Quercus rubra L.

1. Summerwood vessel element, 95X
   a. Simple perforation
2. Summerwood vessel element, 95X
3. Two vasicentric tracheids, 95X, entwined, as in the wood
4. Solitary vasicentric tracheid, 95X
5. Springwood vessel element, 95X
   b. Ray-contact area
6. Springwood vessel element, 95X
   c. Ray-contact area
7. Springwood vessel element, 95X
   d. Ray-contact area
   e. Areas of contact with vasicentric tracheids
8. Springwood vessel element, 95X.
   f. Areas of contact with vasicentric tracheids
9. Fiber, 95X
10. Fiber, 95X

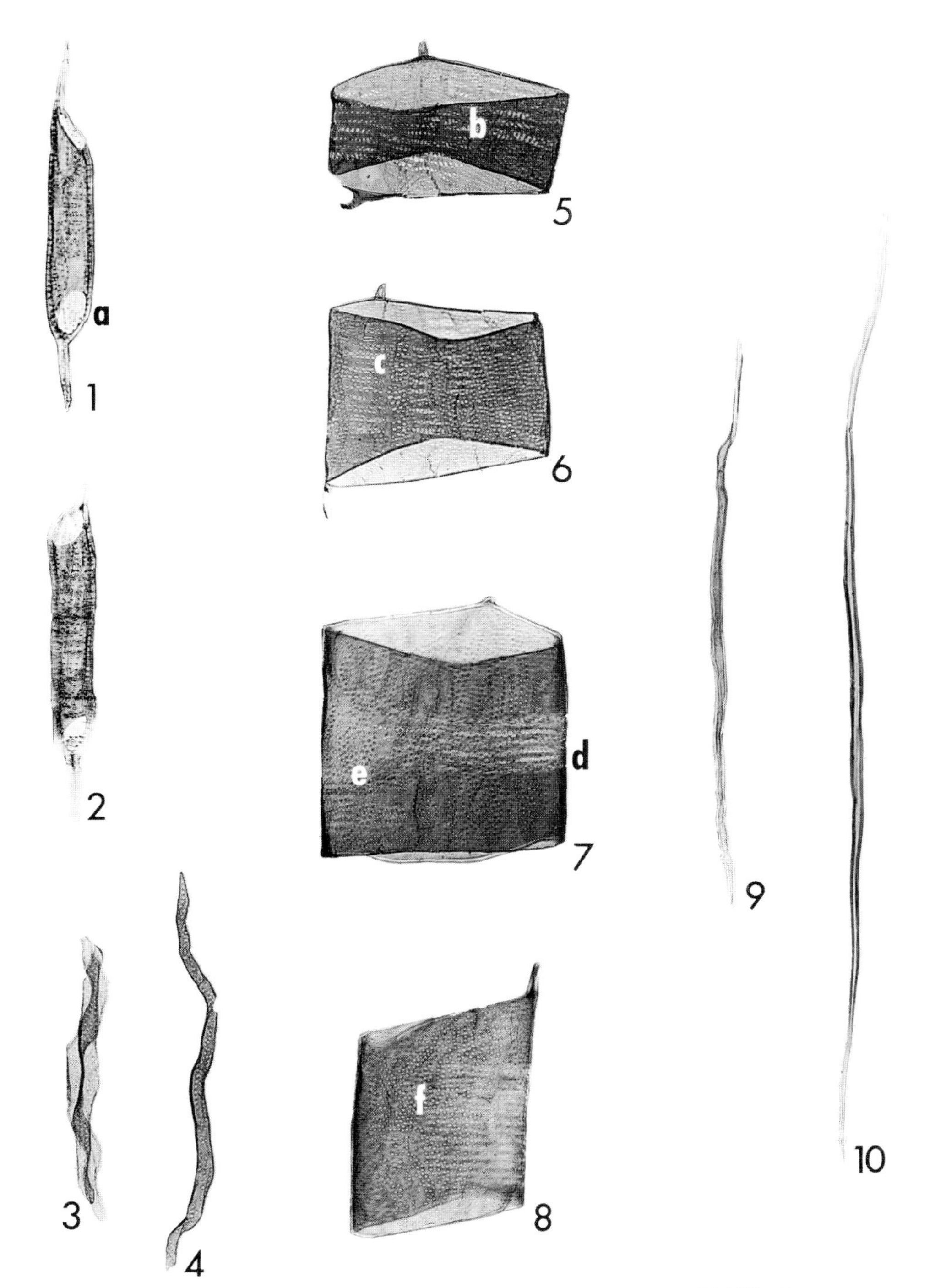

PLATE 41

AMERICAN ELM
Ulmus americana L.

1. Vascular tracheid, 185X, with helical thickening
2. Vascular tracheid, 185X, with helical thickening
3. Summerwood vessel element, 185X, with helical thickening
   a. Simple perforation
4. Vessel element, 185X
   b. Ray-contact area
   c. Area of contact with longitudinal parenchyma
5. Springwood vessel element, 185X. Helical thickening is not evident
   d. Ray-contact area
6. Summerwood vessel element, 185X
   e. Pronounced helical thickening in pitted area
7. Vessel element, 185X, (right) in contact with a narrow vessel element which resembles a vascular tracheid
   f. Simple perforation
   g. Intervessel pitting
8. Vessel element, 185X
9. Short fiber, 185X

PLATE 42

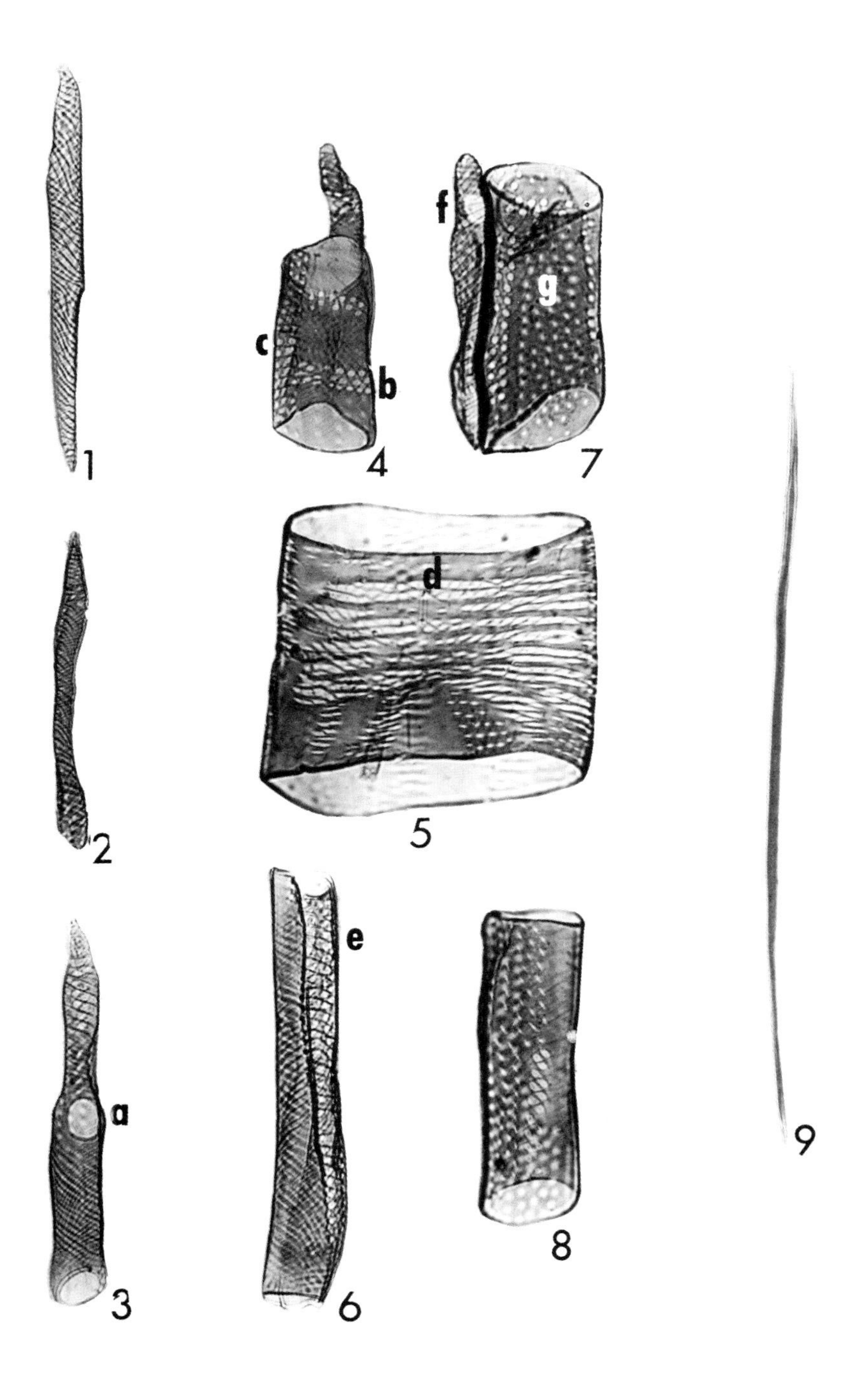
1
2
3
a
4
b
c
5
d
6
e
7
f
g
8
9

CUCUMBERTREE

Magnolia acuminata L.

1. Vessel element, 185X
   - *a.* Scalariform intervessel pitting
   - *b.* Scalariform intervessel pitting
2. Vessel element, 185X
   - *c.* Ray-contact area
3. Vessel element, 185X
   - *d.* Ray-contact area
   - *e.* Simple perforation
4. Fiber, 185X

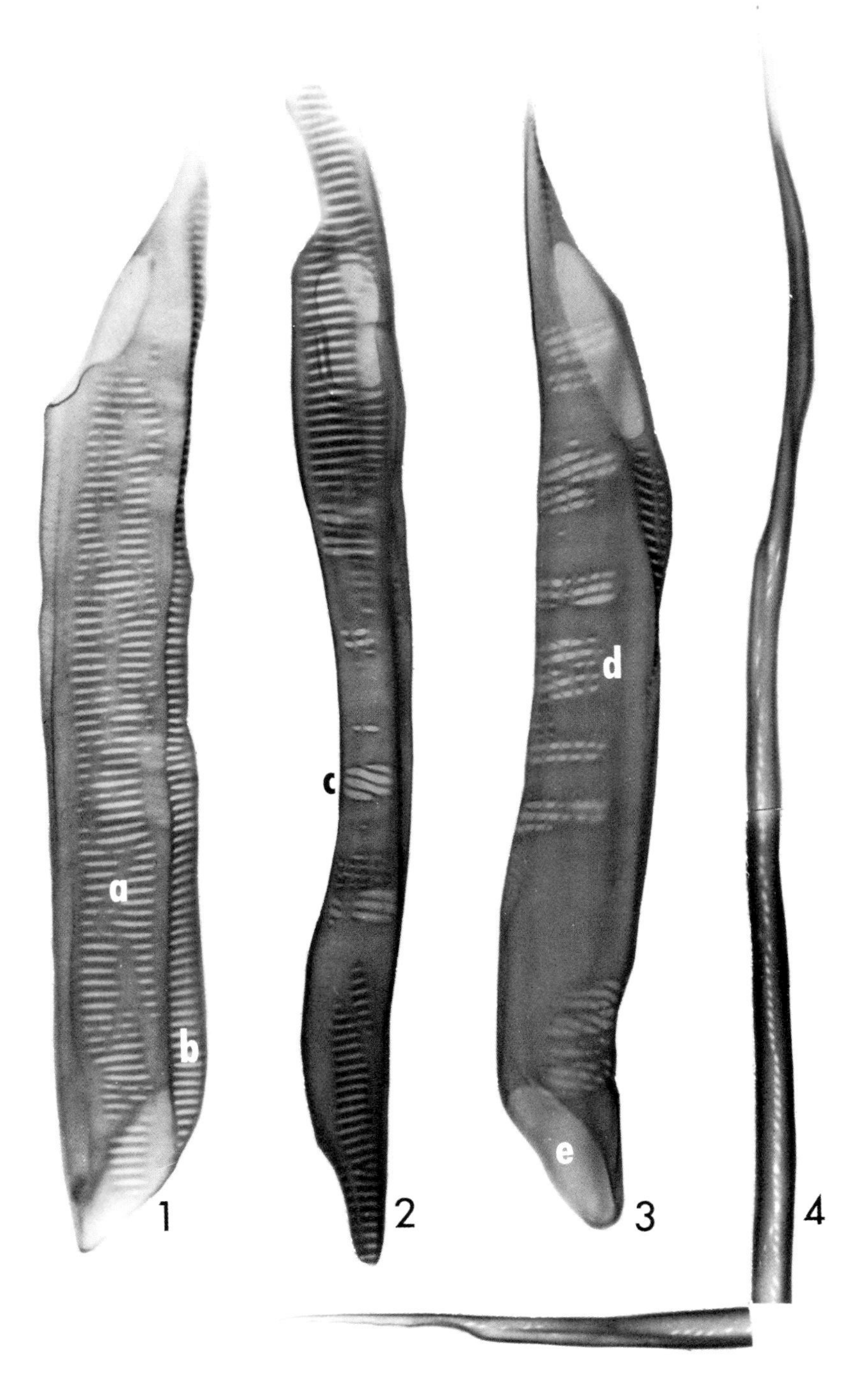
a
b
1
c
2
d
e
3
4

EVERGREEN MAGNOLIA,
SOUTHERN MAGNOLIA
Magnolia grandiflora L.

1. Vessel element, 185X
    - *a.* Scalariform perforation with a few thick bars
    - *b.* Ray-contact area
    - *c.* Helical thickening
    - *d.* Scalariform intervessel pitting
2. Vessel element, 95X
    - e. Scalariform intervessel pitting
3. Narrow vessel element, 95X
4. Narrow vessel element, 95X
5. Fiber, 95X

EVERGREEN MAGNOLIA SOUTHERN MAGNOLIA

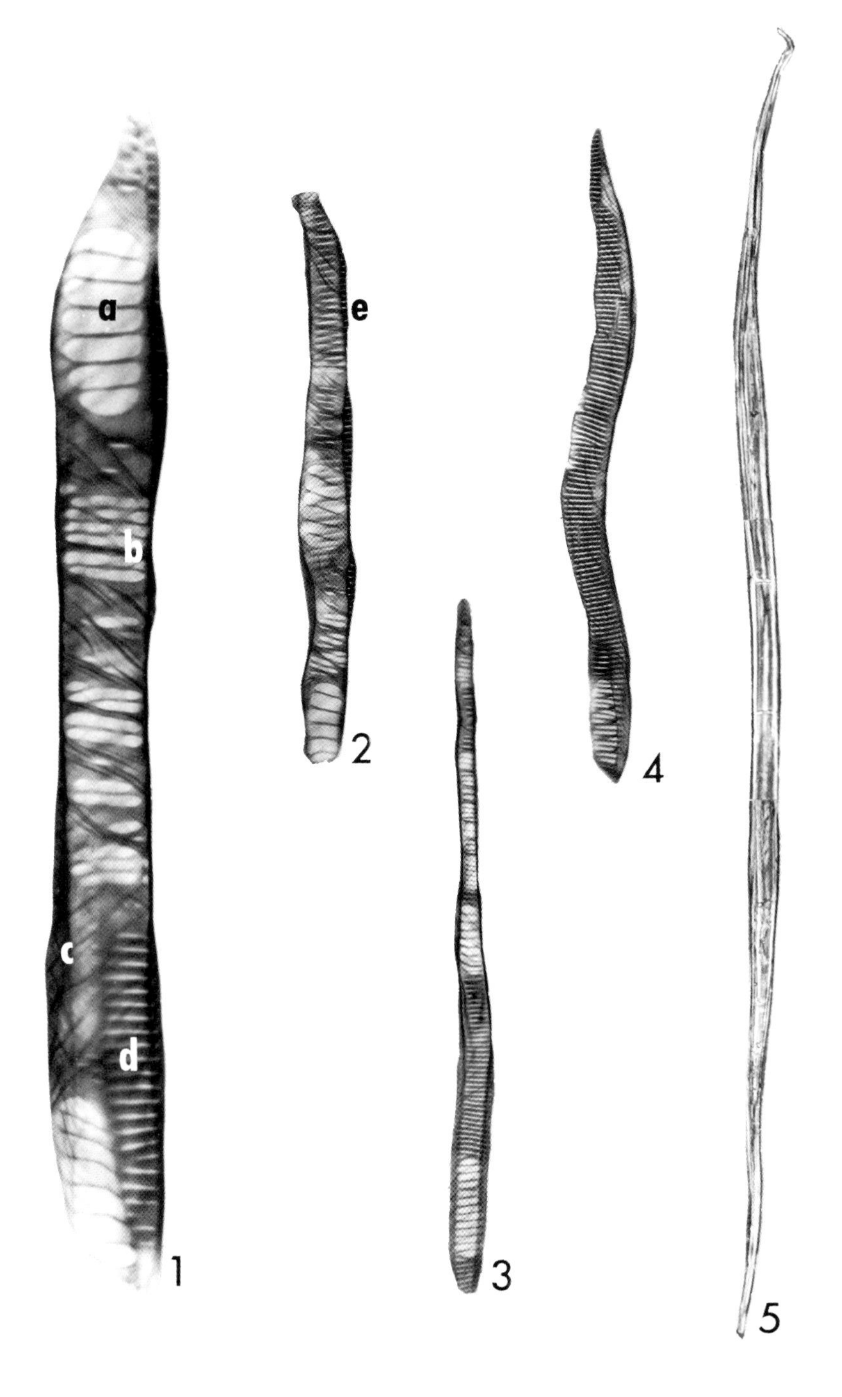

YELLOW-POPLAR, TULIP-POPLAR

Liriodendron tulipifera L.

1. Vessel element, 185X
   - *a.* Scalariform perforation with two stout bars
   - *b.* Intervessel pitting: pits in transverse rows
   - *c.* Ray-contact area
2. Vessel element, 185X
   - *d.* Scalariform perforation with four stout bars
   - *e.* Intervessel pitting: pits in transverse rows
3. Vessel element, 185X
4. Fiber, 185X

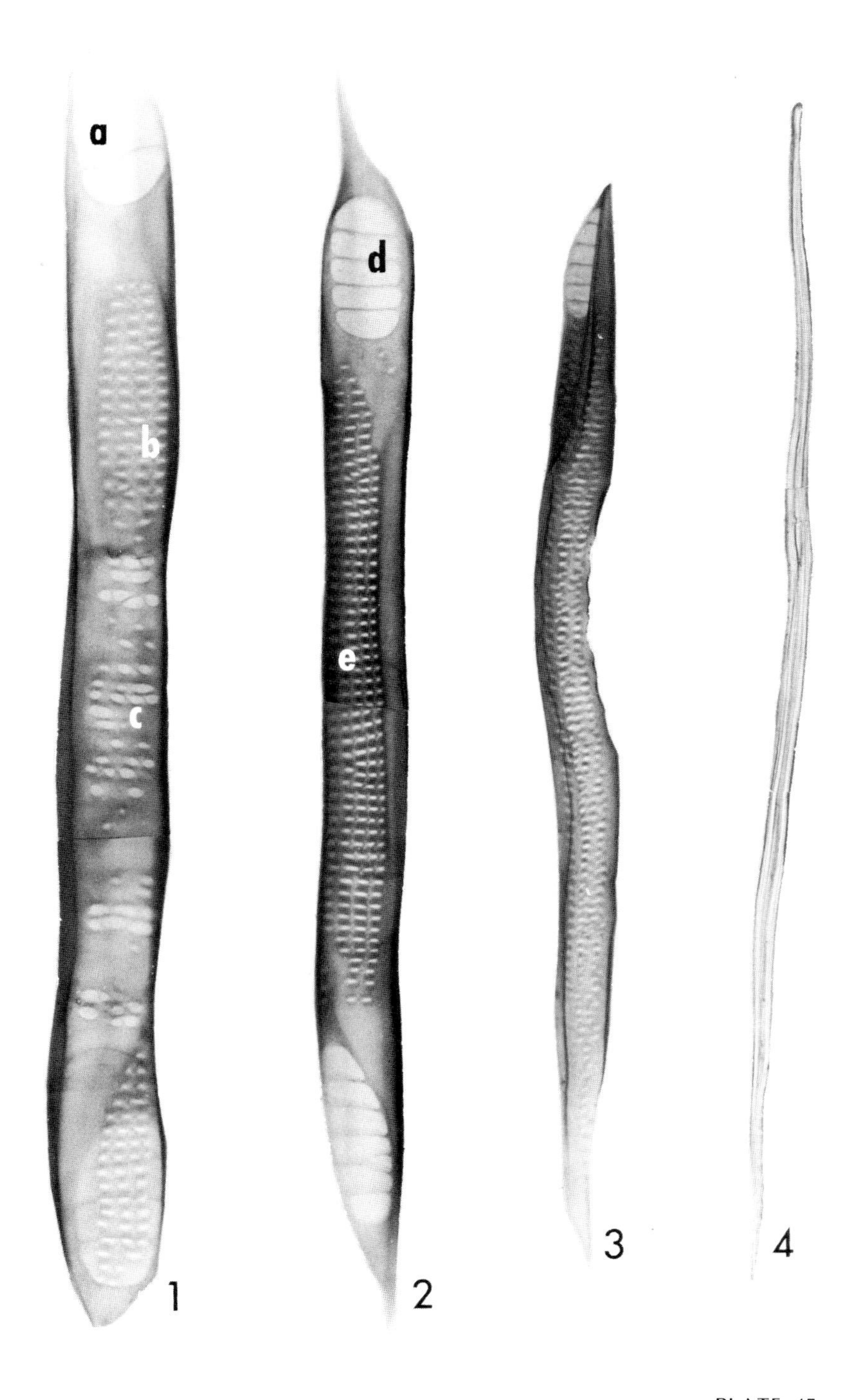
a
b
c
d
e
1
2
3
4

REDBAY

Persea borbonia (L.) Spreng.

1. Vessel element, 95X, with cells of longitudinal parenchyma in contact, as in the wood
   - *a.* Simple perforation in rear wall, visible through the intervessel-pitted front wall
2. Vessel element, 95X
   - *b.* Intervessel pitting: pit apertures confluent
3. Two vessel elements, 95X, in contact, as in the wood
   - *c.* Cells of ray parenchyma in contact, as in the wood
   - *d.* Cells of longitudinal parenchyma in contact, as in the wood
   - *e.* Simple perforation
4. Vessel element, 185X
   - *f.* Ray-contact area
5. Portions of two summerwood vessel elements, 180X, with scalariform perforations at upper ends
6. Fiber, 95X

PLATE 46

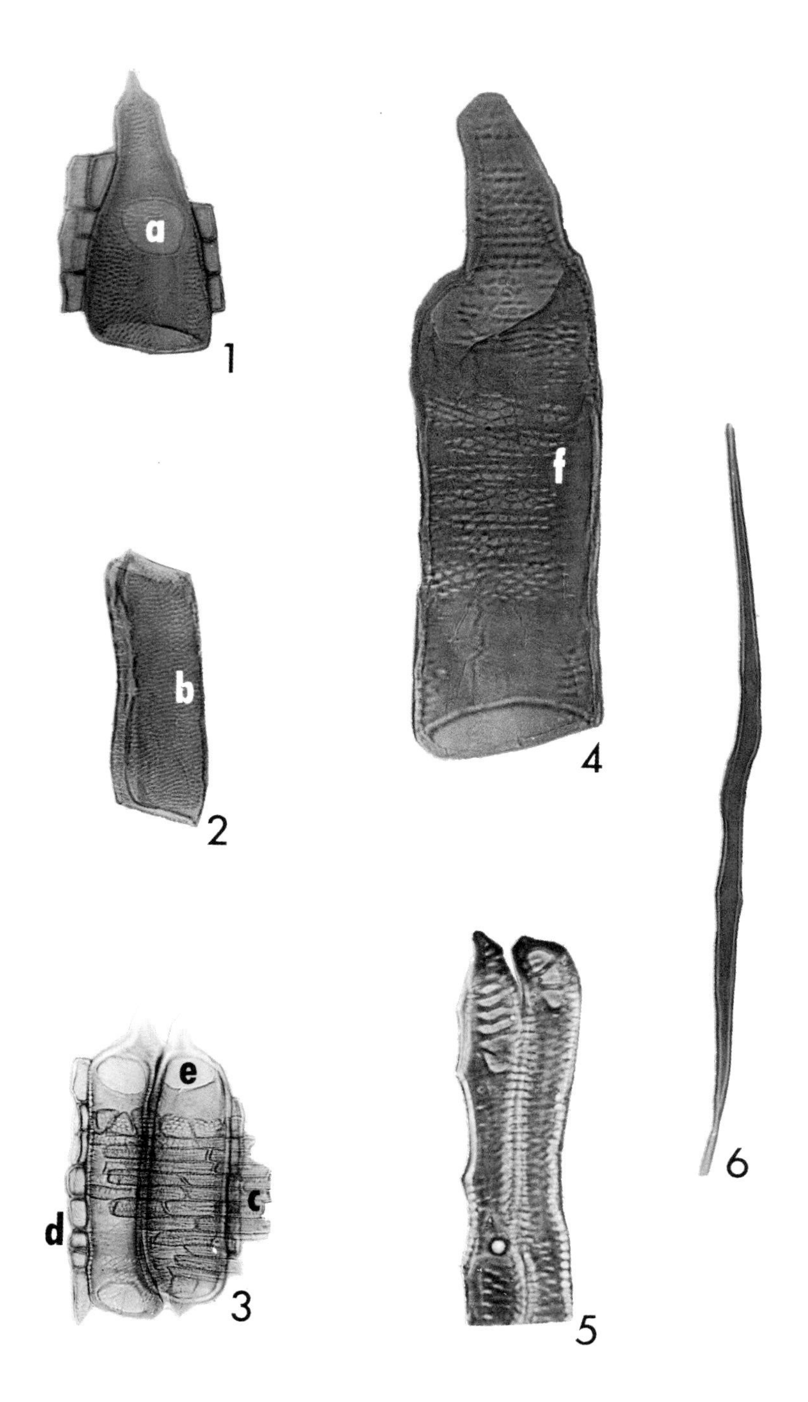

PLATE 46

SASSAFRAS

Sassafras albidum (Nutt.) Nees

1. Summerwood vessel element, 95X
2. Springwood vessel element, 95X, containing a tylosis
   a. Areas of contact with longitudinal parenchyma
3. Springwood vessel element, 185X
   b. Ray-contact area
   c. Fiber-contact area
4. Vessel element, 95X
   d. Intervessel pitting
5. Portion of summerwood vessel element, 95X, with a scalariform perforation with one stout bar
6. Short fiber, 95X
7. Fiber, 95X

PLATE 47

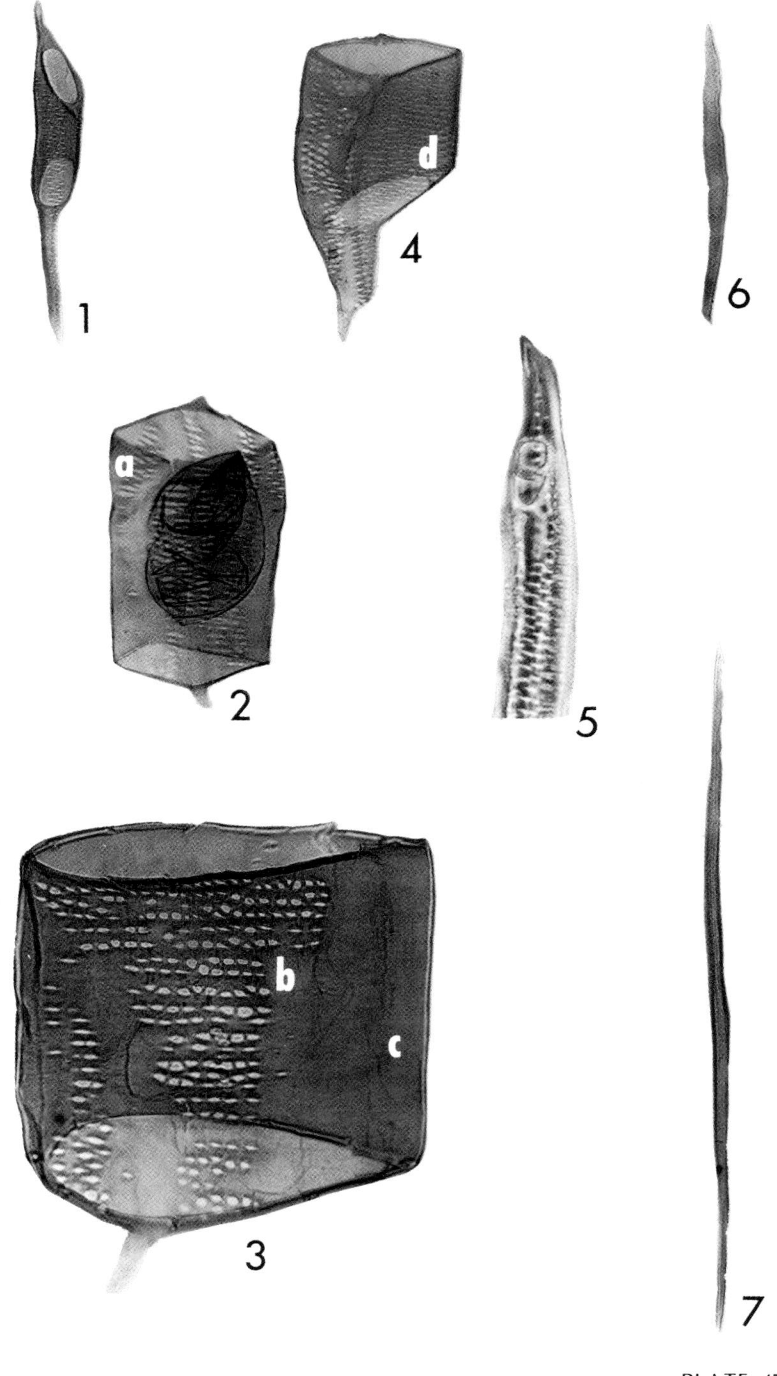

PLATE 47

SWEETGUM, RED-GUM
Liquidambar styraciflua L.

1. Portion of a long vessel element, 185X
    a. Helical thickening confined to the tip
    b. Scalariform perforation with many thin bars
2. Vessel element, 95X
3. Vessel element, 95X
    c. Ray-contact area
    d. Fiber-contact area
4. Vessel element, 95X
    e. Ray-contact area
5. Long vessel element, 95X
6. Fiber, 95X, with bordered pitting

PLATE 48

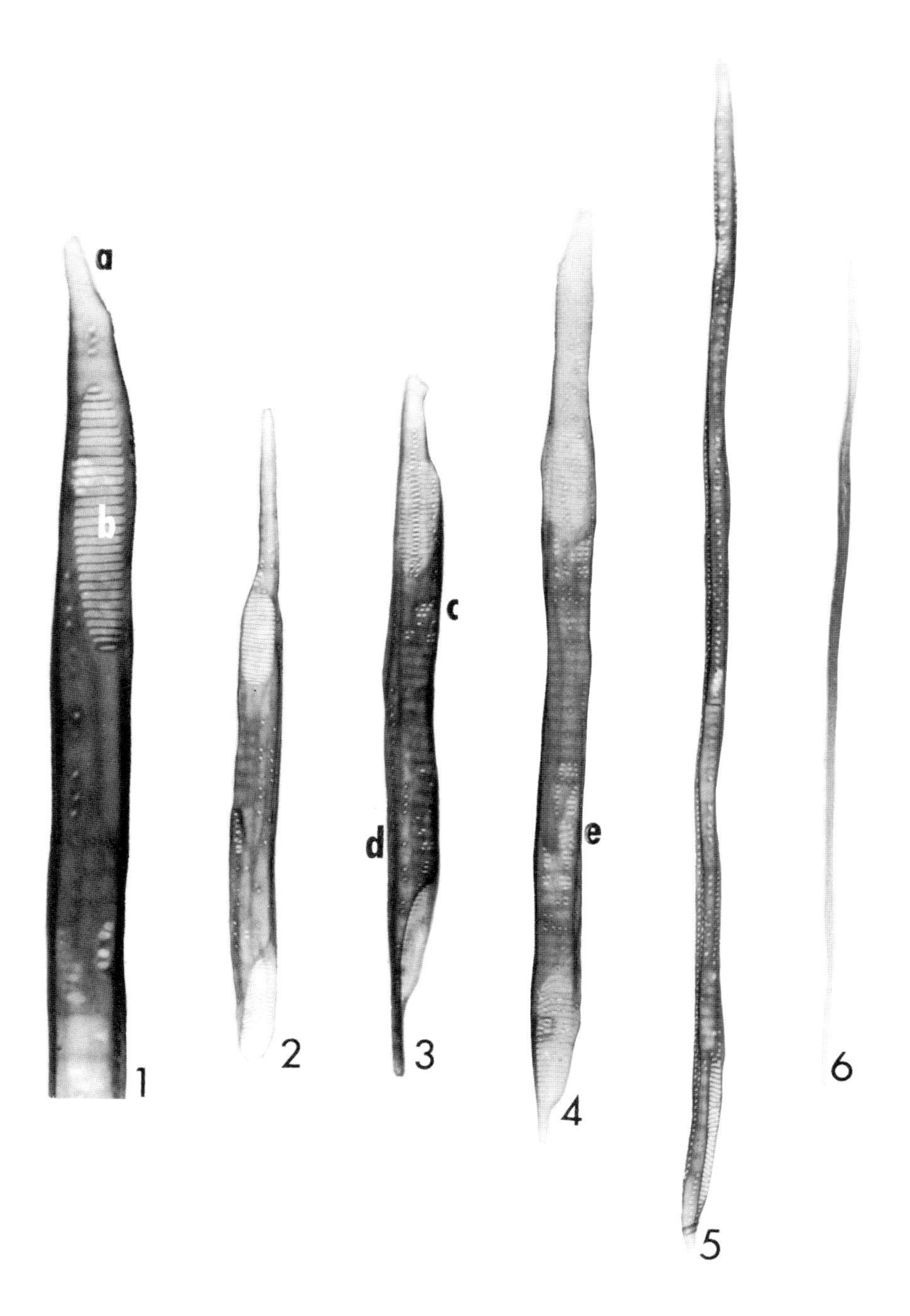
a
b
c
d
e
1
2
3
4
5
6

SYCAMORE

Platanus occidentalis L.

1. Vessel element, 95X
    - *a.* Intervessel pitting
2. Vessel element, 95X
    - *b.* Intervessel pitting
    - *c.* Simple perforation
3. Vessel element, 185X
    - *d.* Scalariform perforation
    - *e.* Intervessel pitting
    - *f.* Area of contact with longitudinal parenchyma
4. Narrow vessel element, 95X
5. Narrow vessel element, 95X
    - *g.* Scalariform perforation with thin bars
    - *h.* Fiber-contact areas
6. Fiber, 95X

PLATE 49

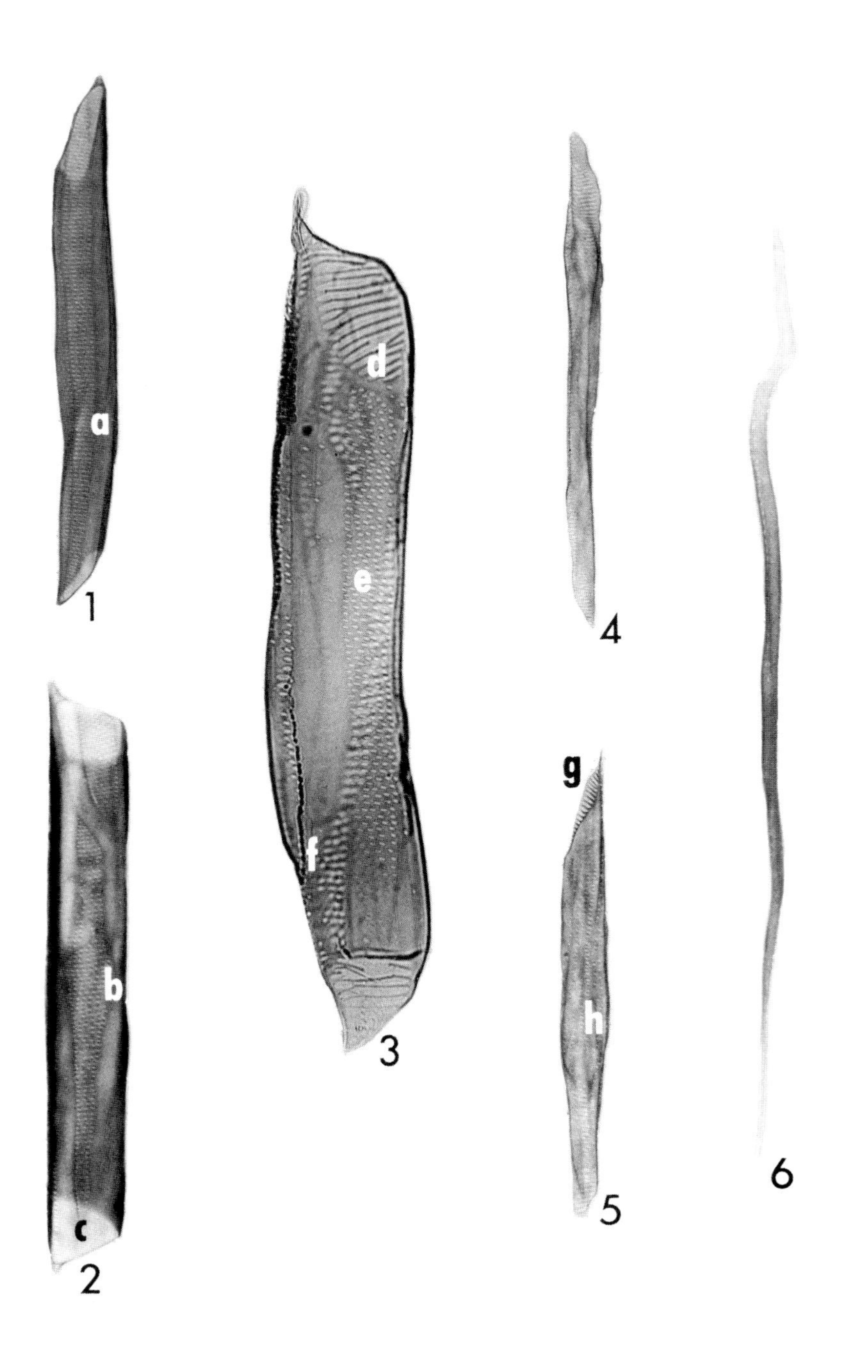
a
1
b
c
2
d
e
f
3
4
g
h
5
6

BLACK CHERRY
Prunus serotina Ehrh.

1. Vessel element, 95X

2. Vessel element, 95X, with fine helical thickening

3. Vessel element, 95X

4. Vessel element, 185X
    - *a.* Fine helical thickening
    - *b.* Ray-contact area
    - *c.* Simple perforation

5. Vessel element, 95X

6. Short fiber, 95X

7. Vessel element, 95X
    - *d.* Intervessel pitting

8. Fiber, 95X

PLATE 50

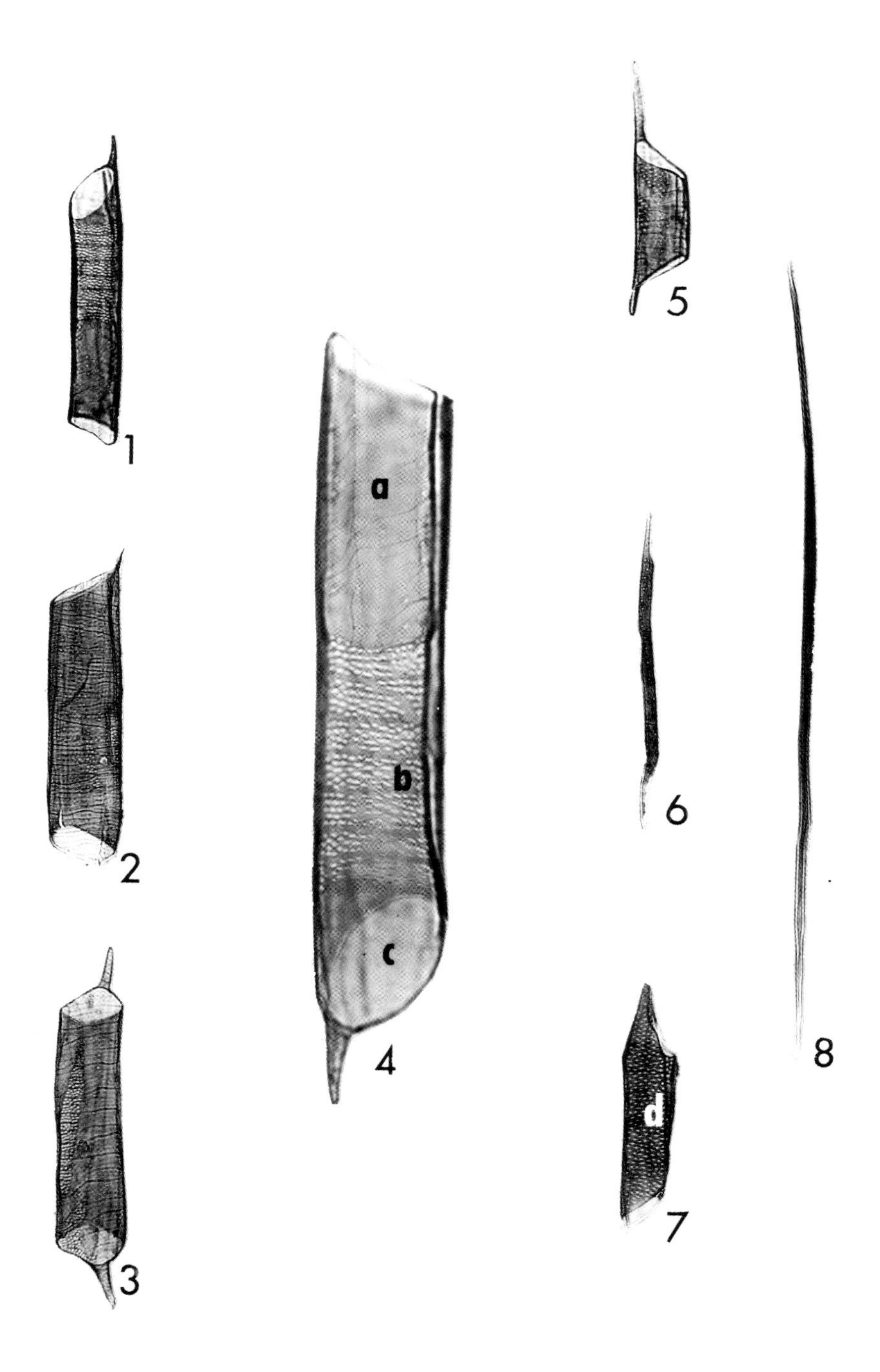
1
2
3
a
b
c
4
5
6
d
7
8

AILANTHUS, TREE-OF-HEAVEN

Ailanthus altissima (Mill.) Swingle

1. Summerwood vessel element, 95X
   - *a*. Simple perforation
   - *b*. Fine helical thickening
2. Summerwood vessel element, 95X, with fine helical thickening
3. Springwood vessel element, 95X
   - *c*. Area of contact with longitudinal parenchyma
4. Summerwood vessel element, 95X
5. Summerwood vessel element, 95X, with fine helical thickening
6. Springwood vessel element, 185X
   - *d*. Simple perforation
   - *e*. Area of contact with longitudinal parenchyma: pit apertures confluent
7. Springwood vessel element, 95X
8. Fiber, 95X
9. Short fiber, 95X

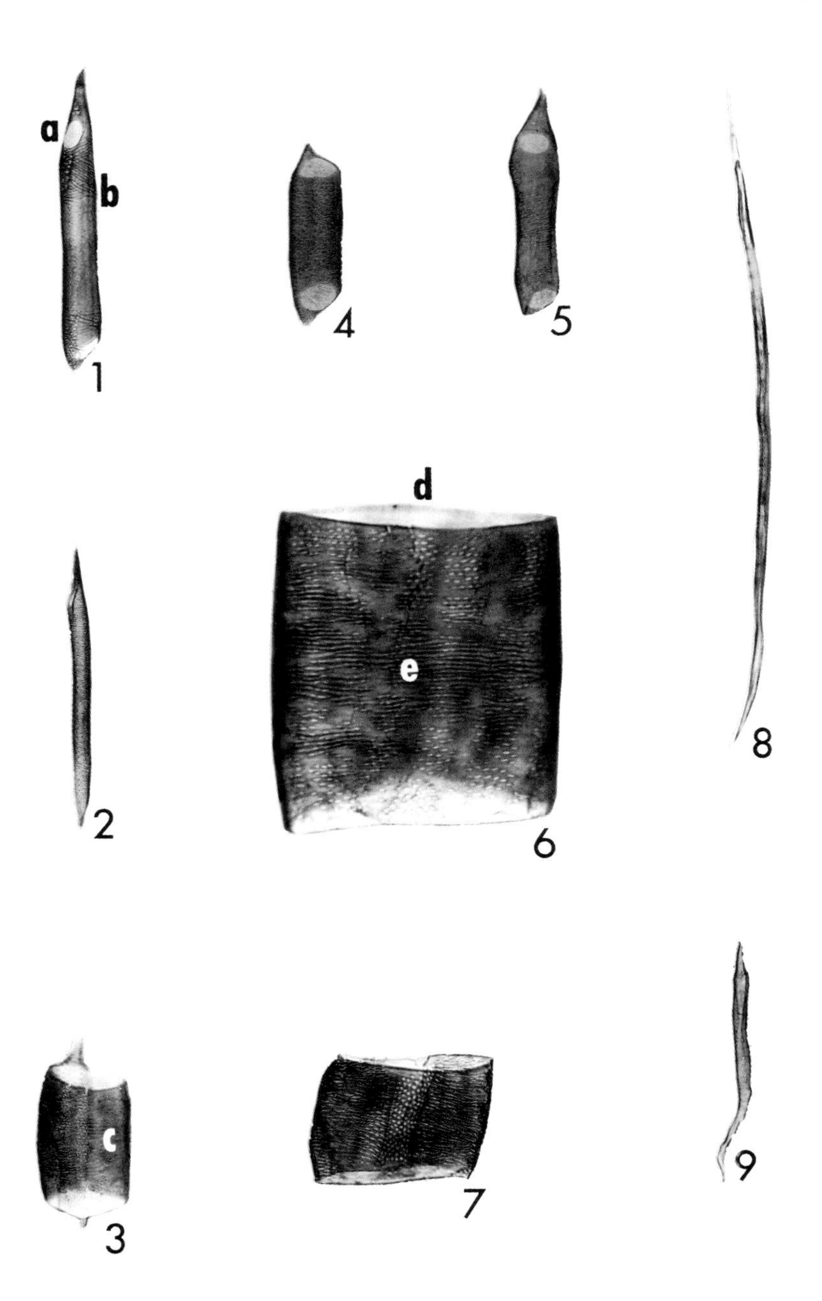
a
b
1
4
5
8
d
e
2
6
c
3
7
9

HOLLY

Ilex opaca Ait.

1. Portion of vessel element, 185X, with helical thickening
   a. Intervessel pitting
2. Vessel element, 95X, with helical thickening
   b. Scalariform perforation
3. Vessel element, 95X, with helical thickening
   c. Scalariform perforation with many thin bars
4. Vessel element, 95X
   d. Helical thickening
5. Fiber, 95X, with helical thickening

PLATE 52

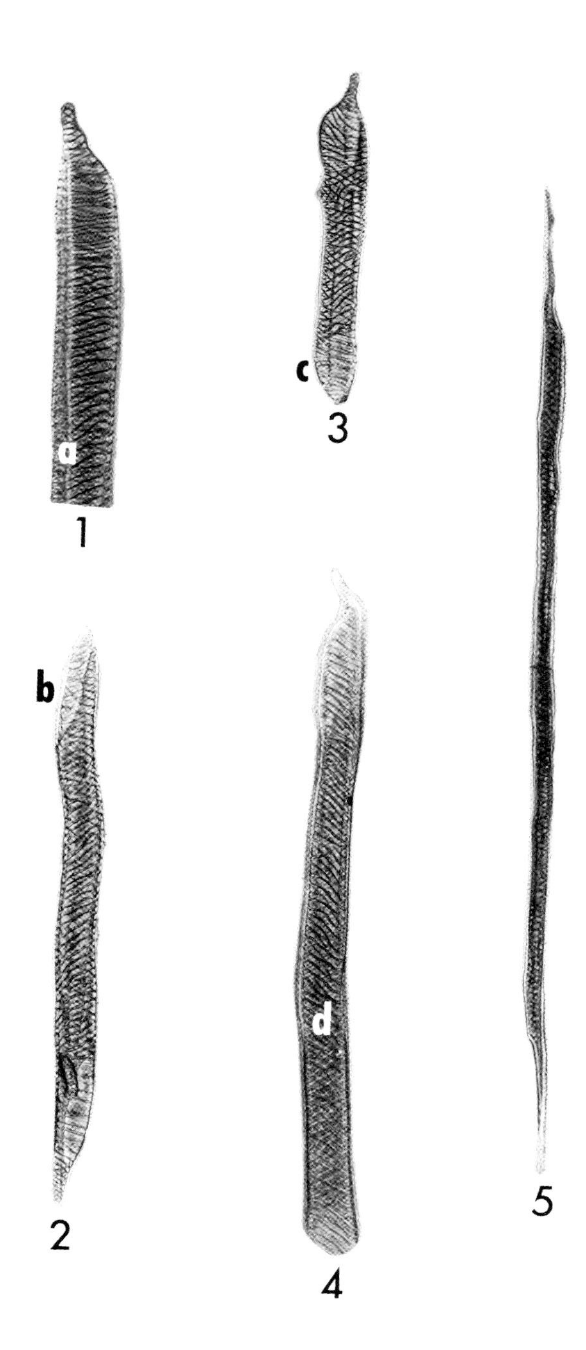
a
1
b
2
c
3
d
4
5

SUGAR MAPLE

Acer saccharum Marsh.

1. Vessel element, 185X
   - *a.* Ray-contact area
   - *b.* Fine helical thickening on wall
2. Vessel element, 185X
   - *c.* Simple perforation
3. Vessel element, 185X
   - *d.* Intervessel pitting
4. Vessel element, 185X
   - *e.* Helical thickening
5. Vessel element, 185X
   - *f.* Areas of contact with longitudinal parenchyma
6. Fiber, 185X

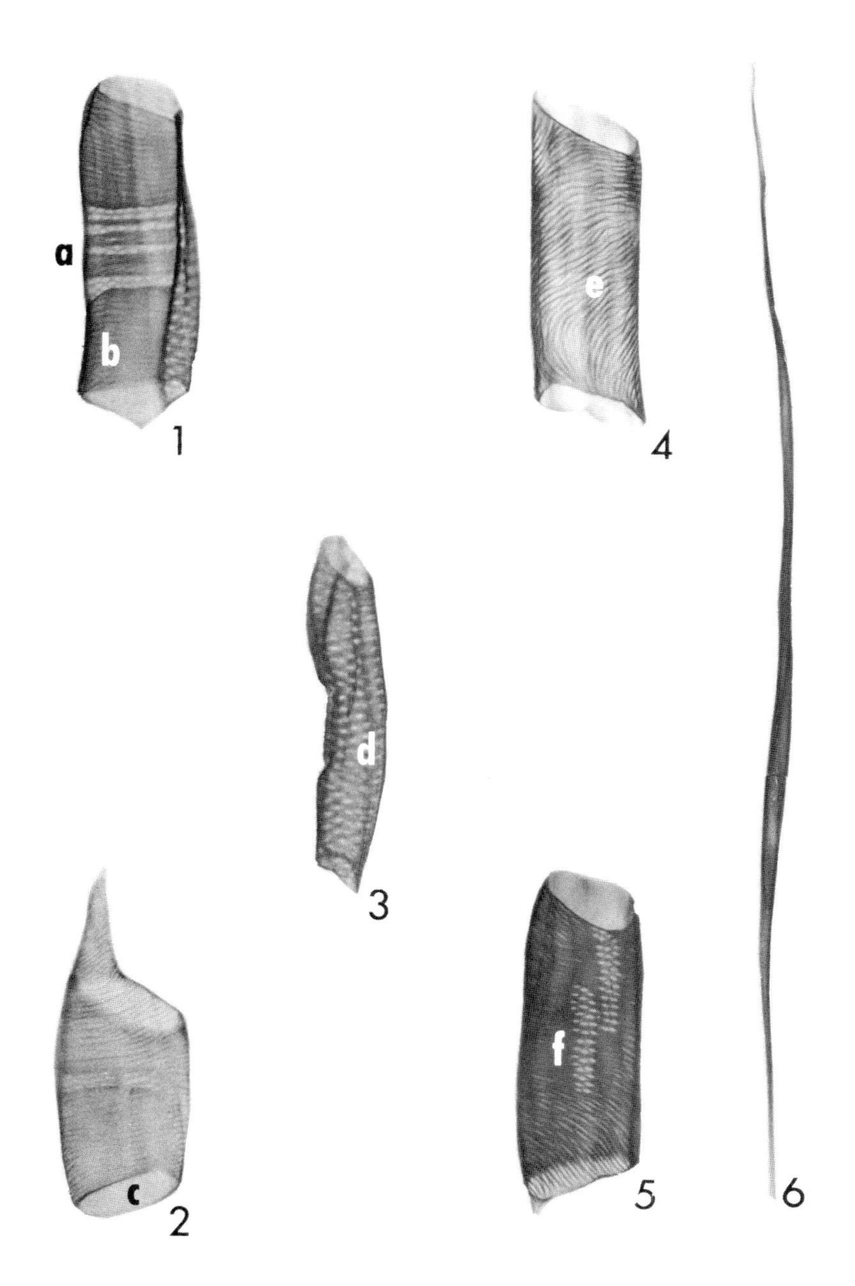
a
b
1
e
4
d
3
c
2
f
5
6

RED MAPLE
Acer rubrum L.

1. Vessel element, 185X, with helical thickening
    a. Simple perforation
    b. Area of contact with longitudinal parenchyma
    c. Ray-contact area
2. Vessel element, 185X, with helical thickening
    d. Cells of longitudinal parenchyma in contact, as in the wood
3. Vessel element, 185X, with helical thickening
    e. Cells of ray parenchyma in contact, as in the wood
4. Vessel element, 185X, with helical thickening
5. Fiber, 185X

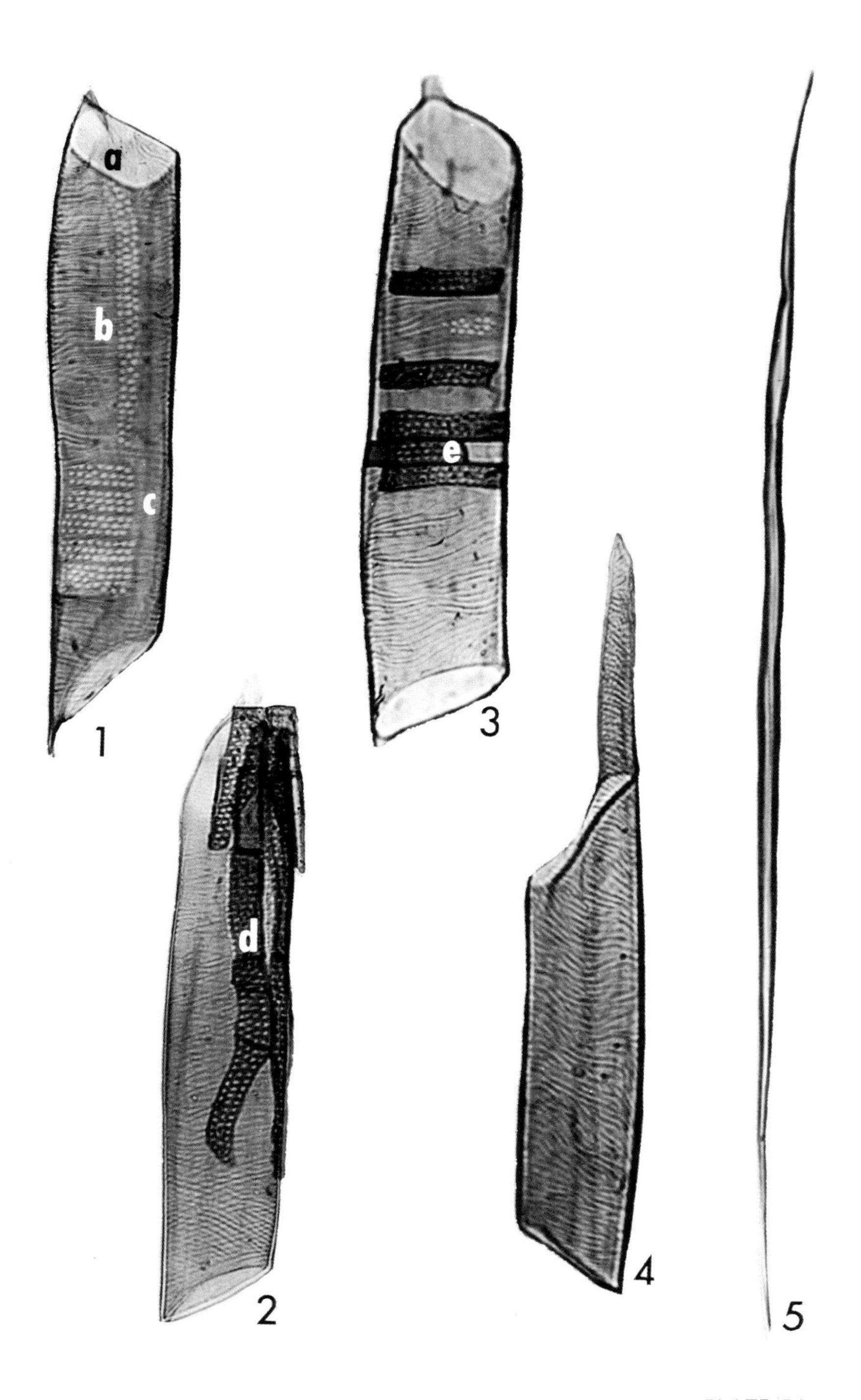
a
b
c
1
d
2
e
3
4
5

SILVER MAPLE

Acer saccharinum L.

1. Vessel element, 185X
   - *a.* Intervessel pitting
   - *b.* Ray-contact area
2. Vessel element, 185X
   - *c.* Intervessel pitting
3. Vessel element, 185X
   - *d.* Helical thickening
   - e. Ray-contact area
   - *f.* Simple perforation
4. Vessel element, 185X, with helical thickening
   - *g.* Areas of contact with longitudinal parenchyma
5. Vessel element, 185X, with helical thickening
6. Fiber, 185X

PLATE 55

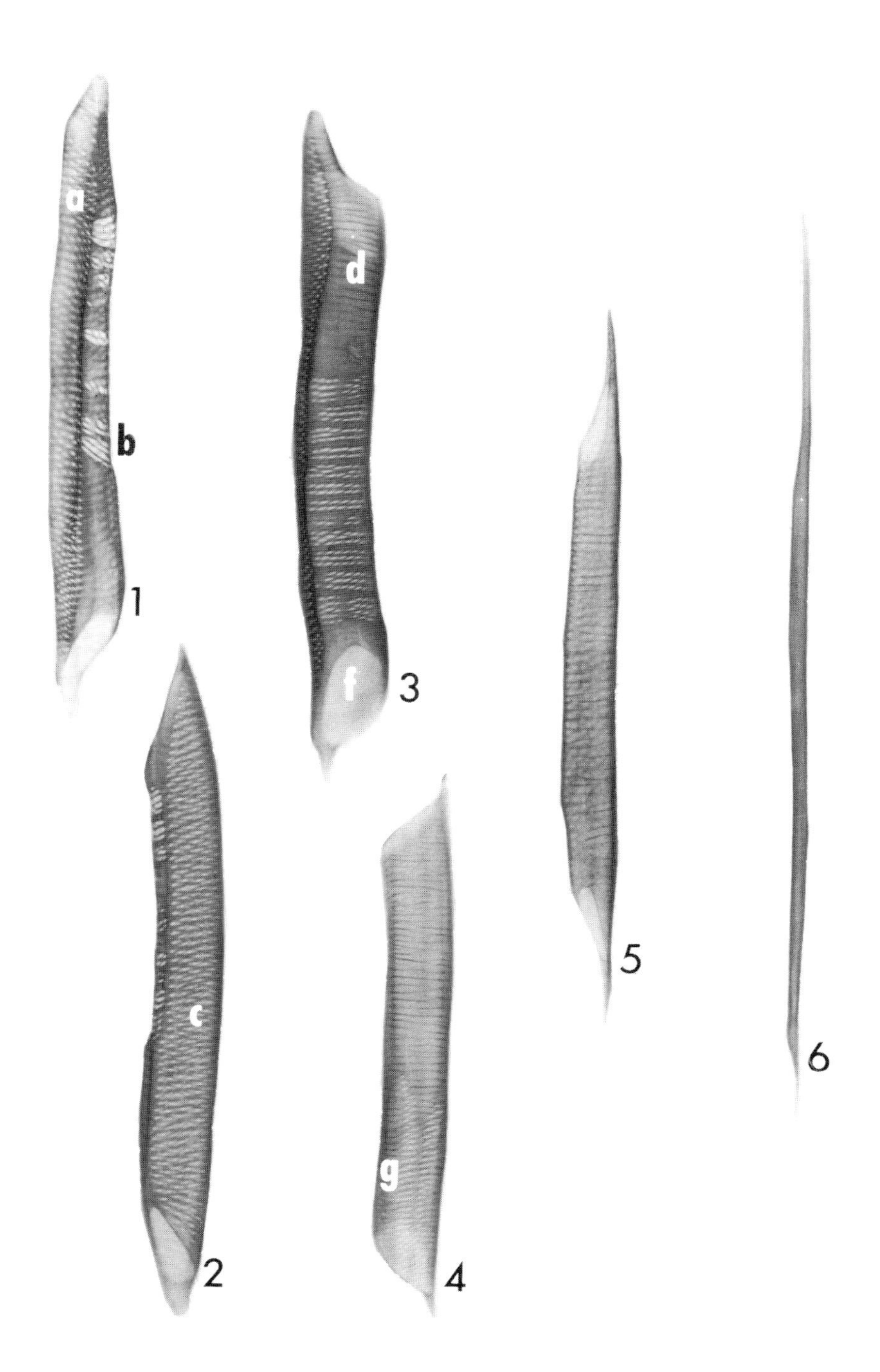
a
b
1
d
f
3
c
2
g
4
5
6

YELLOW BUCKEYE
Aesculus octandra Marsh.

1. Vessel element, 185X
   a. Ray-contact area
   b. Helical thickening
   c. Simple perforation

2. Vessel element, 185X
   d. Intervessel pitting

3. Vessel element, 185X
   e. Helical thickening
   f. Ray-contact area

4. Vessel element, 185X

5. Fiber, 185X

PLATE 56

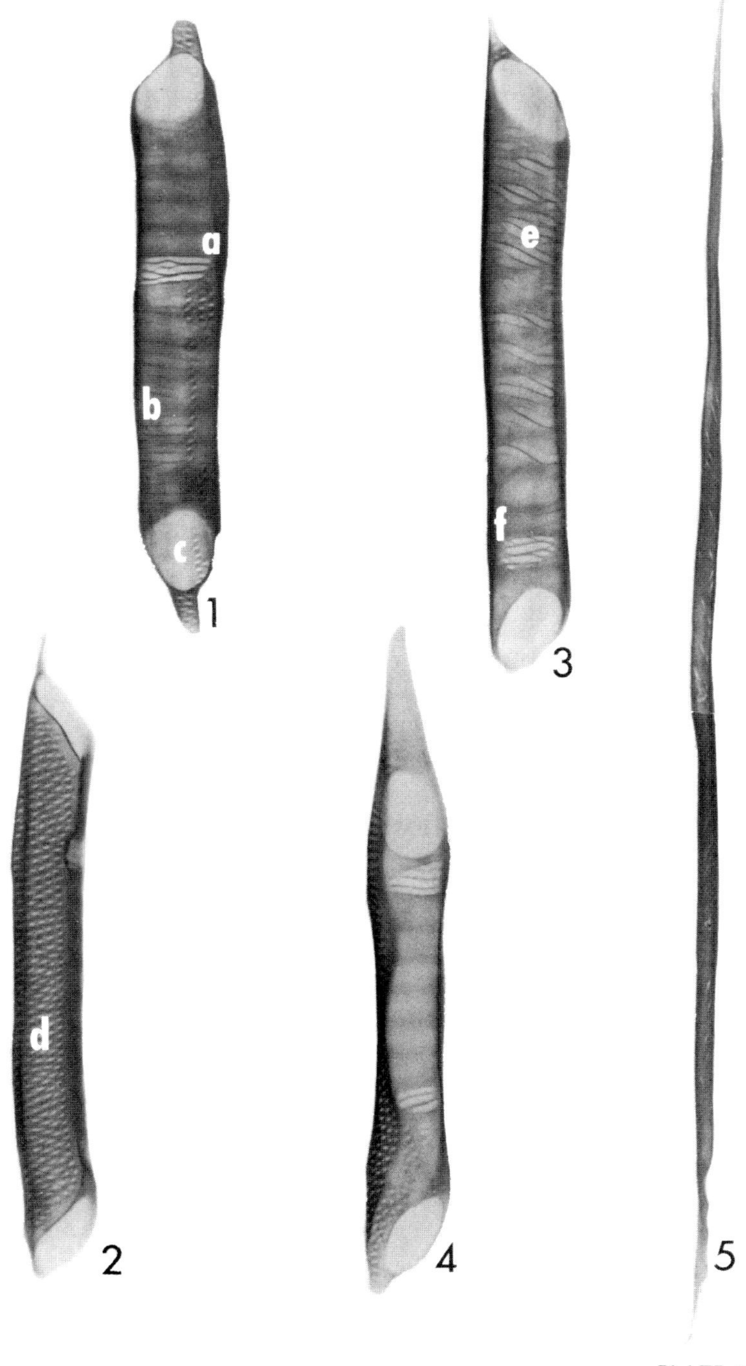
a
b
c
1
d
2
e
f
3
4
5

BASSWOOD

Tilia americana L.

1. Vessel element, 185X
   - *a.* Helical thickening
2. Vessel element, 185X, with spiral thickening
   - *b.* Simple perforation
   - *c.* Ray-contact area, with helical thickening on rear wall
3. Two vessel elements, 185X, in contact, as in the wood
   - *d.* Intervessel pitting and helical thickening
4. Fiber, 185X

PLATE 57

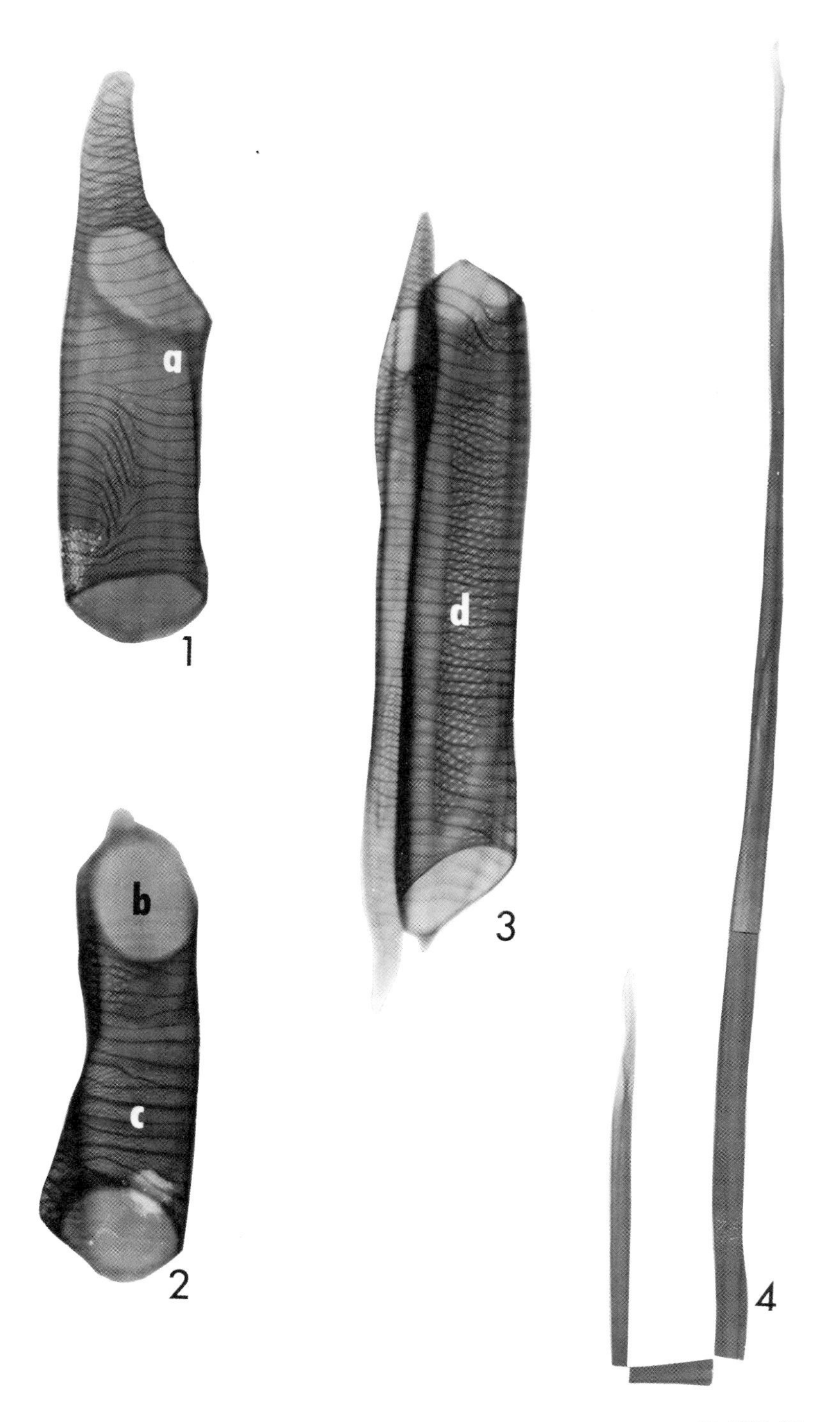
a
1
b
c
2
d
3
4

LOBLOLLY-BAY

Gordonia lasianthus (L.) Ellis

1. Vessel elements, 95X
   - *a.* Scalariform perforation
   - *b.* Ray-contact areas
2. Portion of vessel element, 285X, with helical thickening in tip
3. Vessel element, 95X, with conspicuous tyloses
4. Vessel element, 95X
   - *c.* Helical thickening in tip
   - *d.* Scalariform perforation
   - *e.* Fiber-contact area
5. Fiber, 95X, with bordered pitting
6. Fiber, 95X, with bordered pitting

PLATE 58

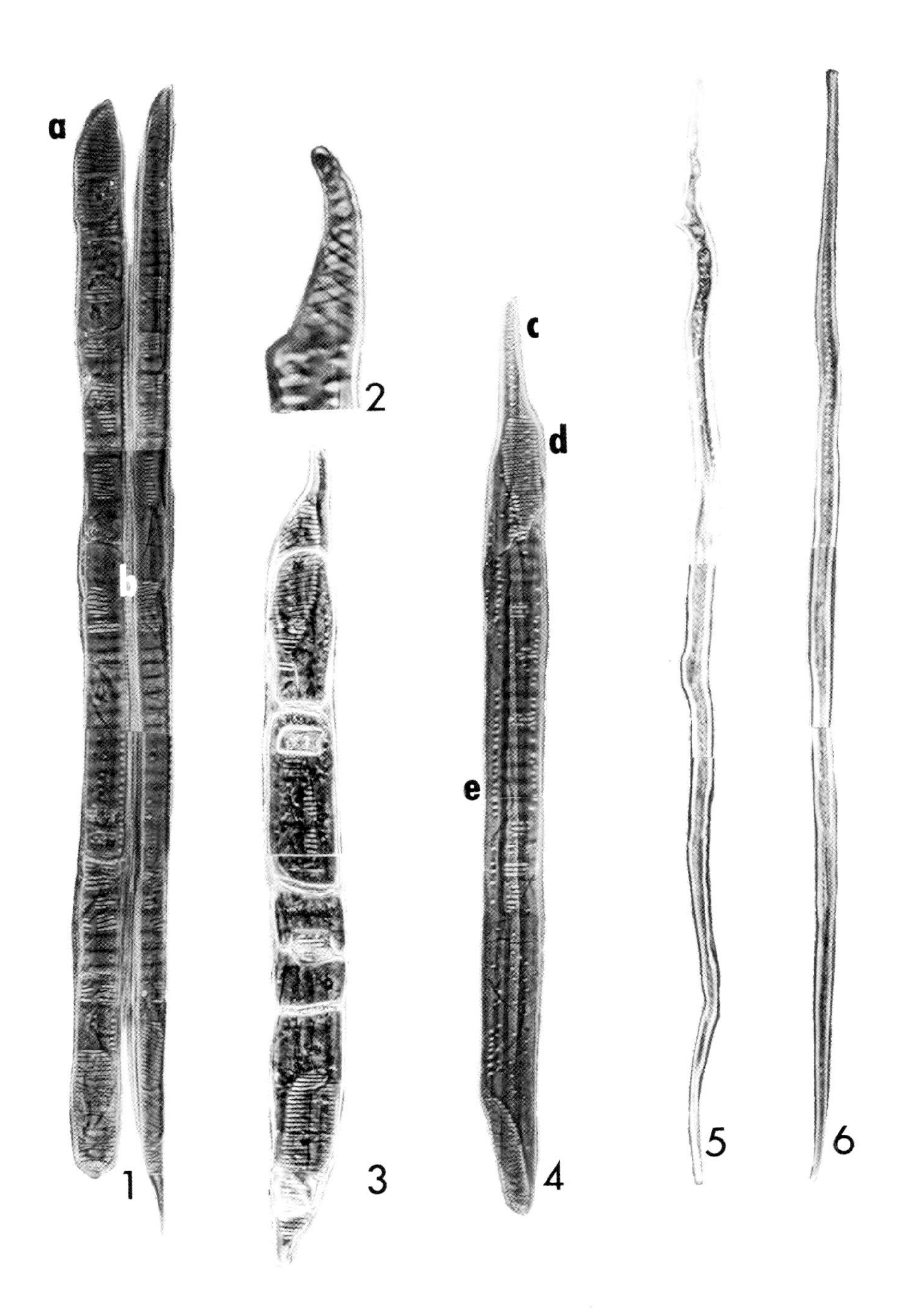

PLATE 58

KAMARERE
Eucalyptus deglupta Blume

1. Vessel element, 95X
    - *a*. Ray-contact area
    - *b*. Simple perforation
    - *c*. Intervessel pitting
2. Vessel element, 95X
    - *d*. Area in contact with vasicentric tracheids
3. Vessel element, 95X
4. Fibers, 95X
5. Vasicentric tracheids, 95X

PLATE 59

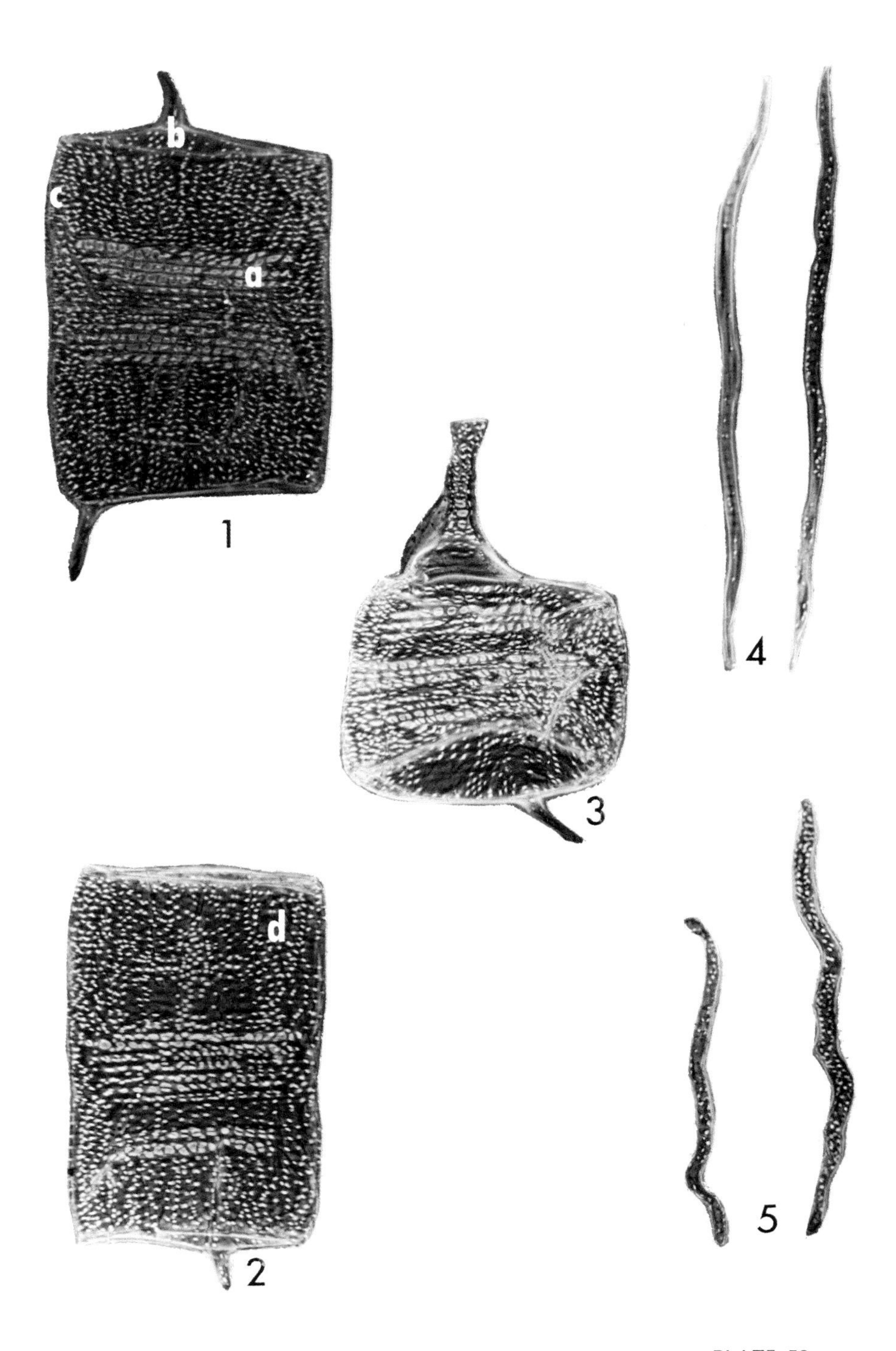

PLATE 59

BLUE GUM

Eucalyptus globulus Labill.

1. Vessel element, 95X
   a. Ray-contact area
   b. Intervessel pitting
2. Vessel element, 95X
   c. Vasicentric tracheid
   d. Simple perforation
3. Vessel element, 95X
4. Small vessel element, 95X, with ligulate tip
5. Fibers, 95X
6. Vasicentric tracheids, 95X
7. Fiber, 95X

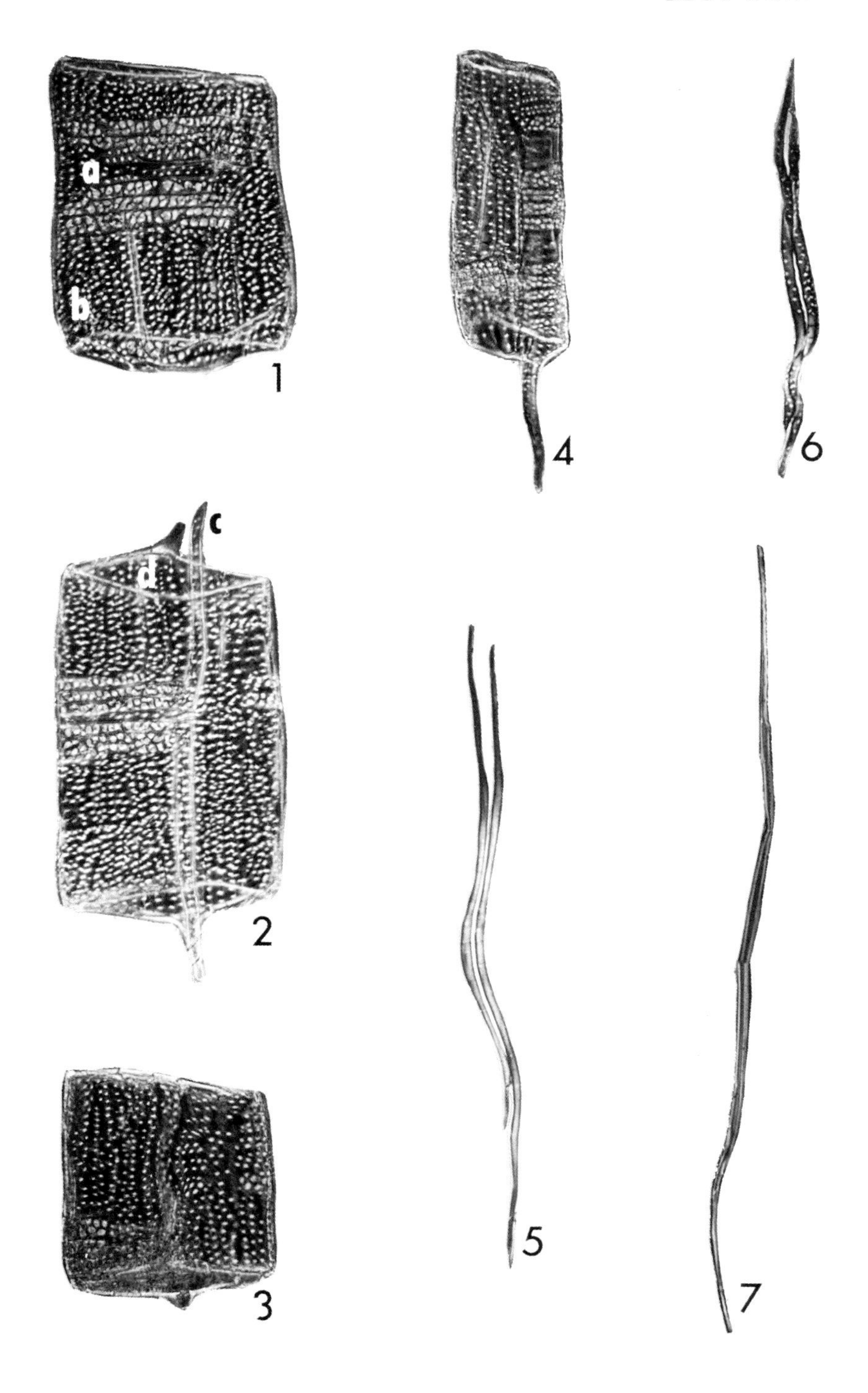

PLATE 60

WATER TUPELO, TUPELO-GUM
Nyssa aquatica L.

1. Vessel element, 185X
   - *a.* Scalariform intervessel pitting
   - *b.* Ray-contact area
   - *c.* Scalariform perforation with many fine bars
2. Vessel element, 95X
   - *d.* Ray-contact area
3. Vessel element, 95X
   - e. Fiber-contact area
   - *f.* Intervessel pitting: fine pits in transverse rows
4. Portion of vessel element, 85X, with helical thickening in tip
5. Wide-lumened fiber, 95X
6. Narrow-lumened fiber, 95X

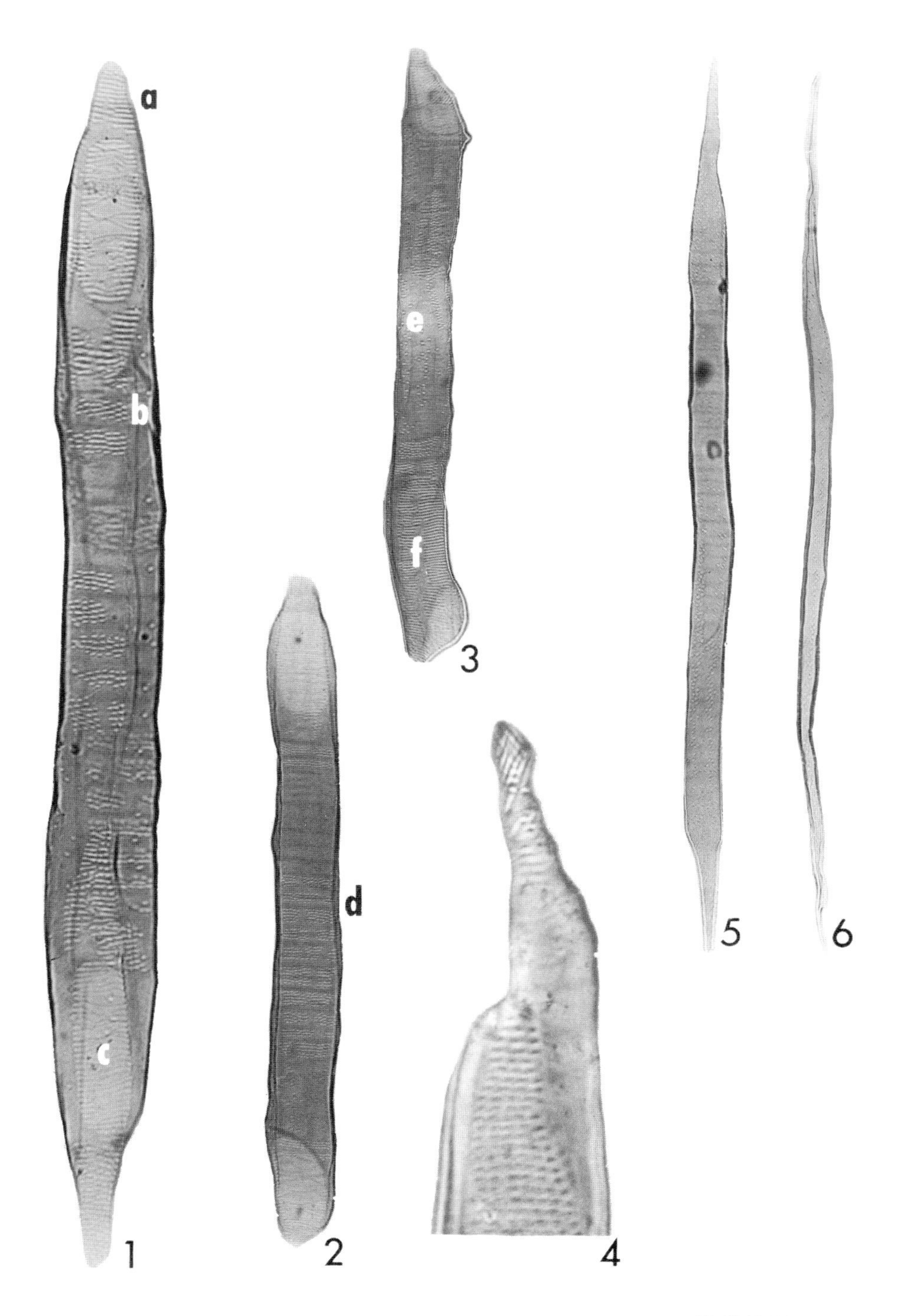

PLATE 61

BLACK TUPELO, BLACKGUM
Nyssa sylvatica Marsh.

1. Portion of a long vessel element, 185X
   - *a.* Scalariform perforation with many fine bars
   - *b.* Ray-contact area
2. Vessel element, 95X
   - *c.* Intervessel pitting: fine pits in transverse rows
   - *d.* Fiber-contact area
3. Portion of vessel element, 185X, with helical thickening in tip
4. Narrow vessel element, 95X
   - e. Intervessel pitting: fine pits in transverse rows
5. Fiber, 95X

PLATE 62

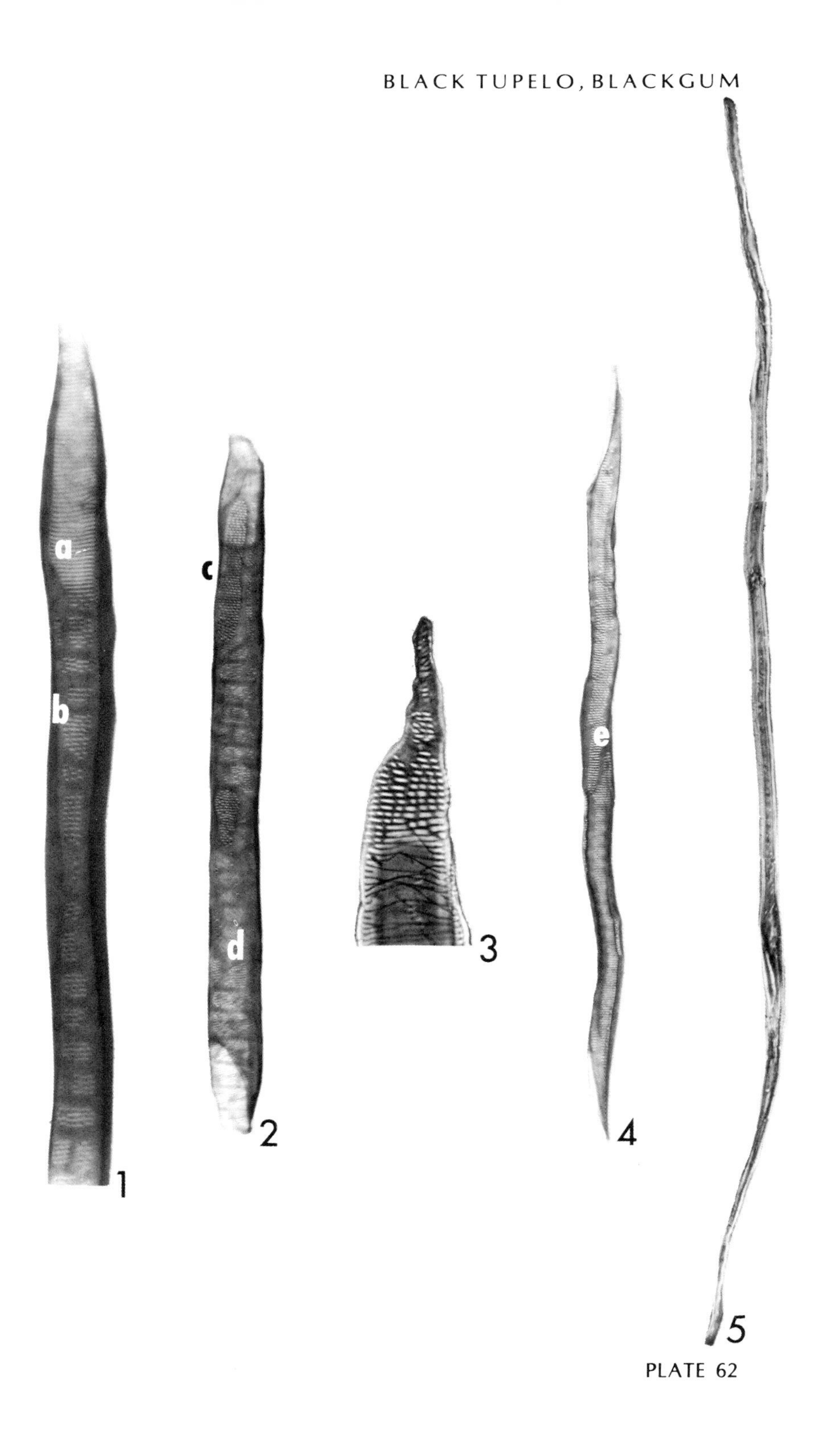

PLATE 62

WHITE ASH

Fraxinus americana L.

1. Springwood vessel element, 95X
   a. Intervessel pitting
2. Summerwood vessel element, 95X
3. Springwood vessel element, 95X
4. Springwood vessel element, 95X
   b. Ray-contact area
   c. Simple perforation
5. Two vessel elements, 95X, in contact, as in the wood
   d. Simple perforation
   e. Area of contact with longitudinal parenchyma
6. Vessel element, 95X
7. Summerwood vessel element, 95X
8. Fiber, 95X

PLATE 63

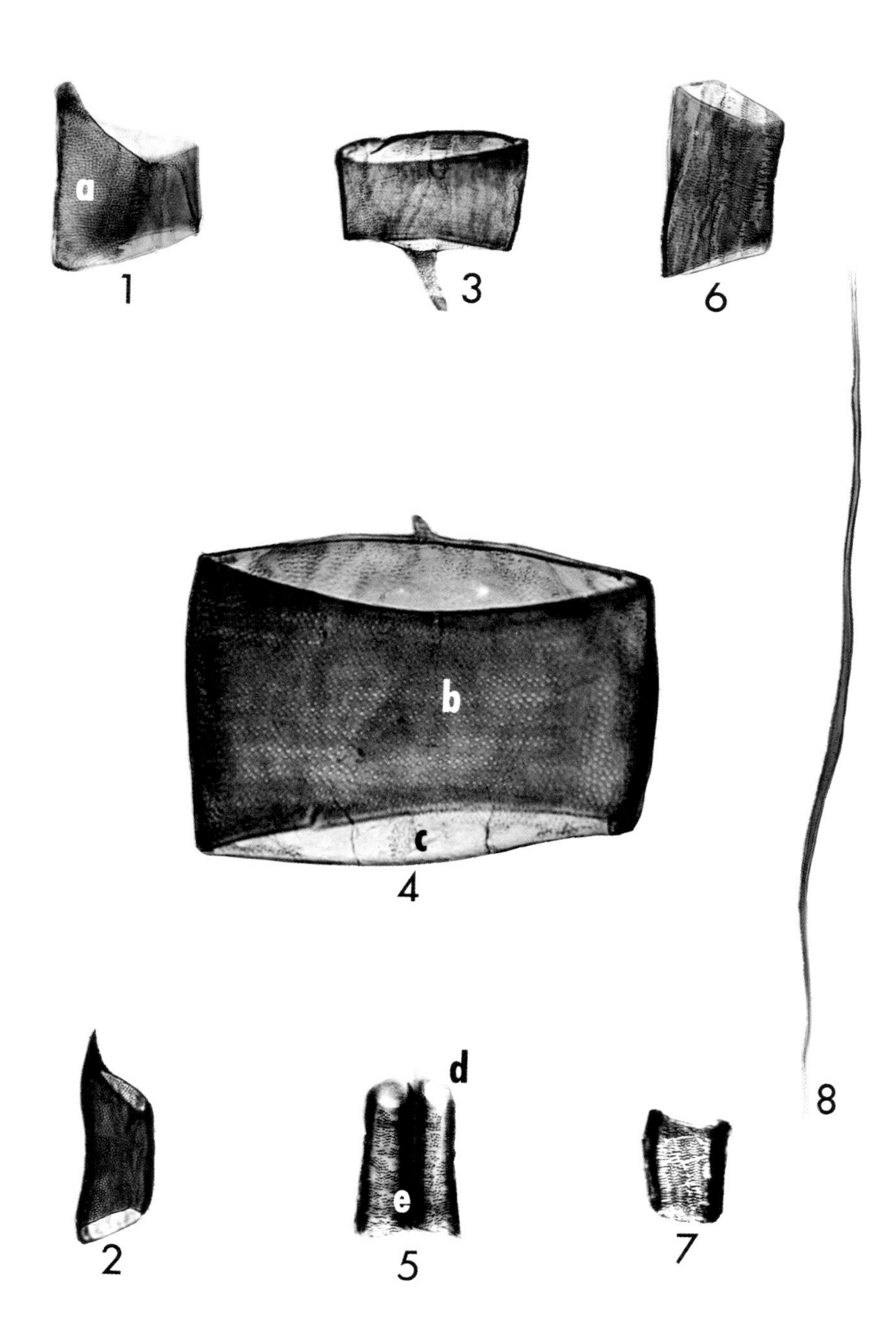
a
1
3
6
b
c
4
8
d
e
2
5
7

NORTHERN CATALPA
Catalpa speciosa Warder

1. Vessel element, 185X
    - *a.* Intervessel pitting
    - *b.* Simple perforation
2. Vessel element, 185X
    - *c.* Area of contact with longitudinal parenchyma
3. Summerwood vessel element, 185X
    - *d.* Areas of contact with longitudinal parenchyma
4. Springwood vessel element, 185X
    - *e.* Ray-contact area
5. Narrow, summerwood vessel element, 185X
6. Short, wide-lumened fiber, 185X
7. Summerwood vessel element, 185X, with helical thickening
8. Fiber, 185X

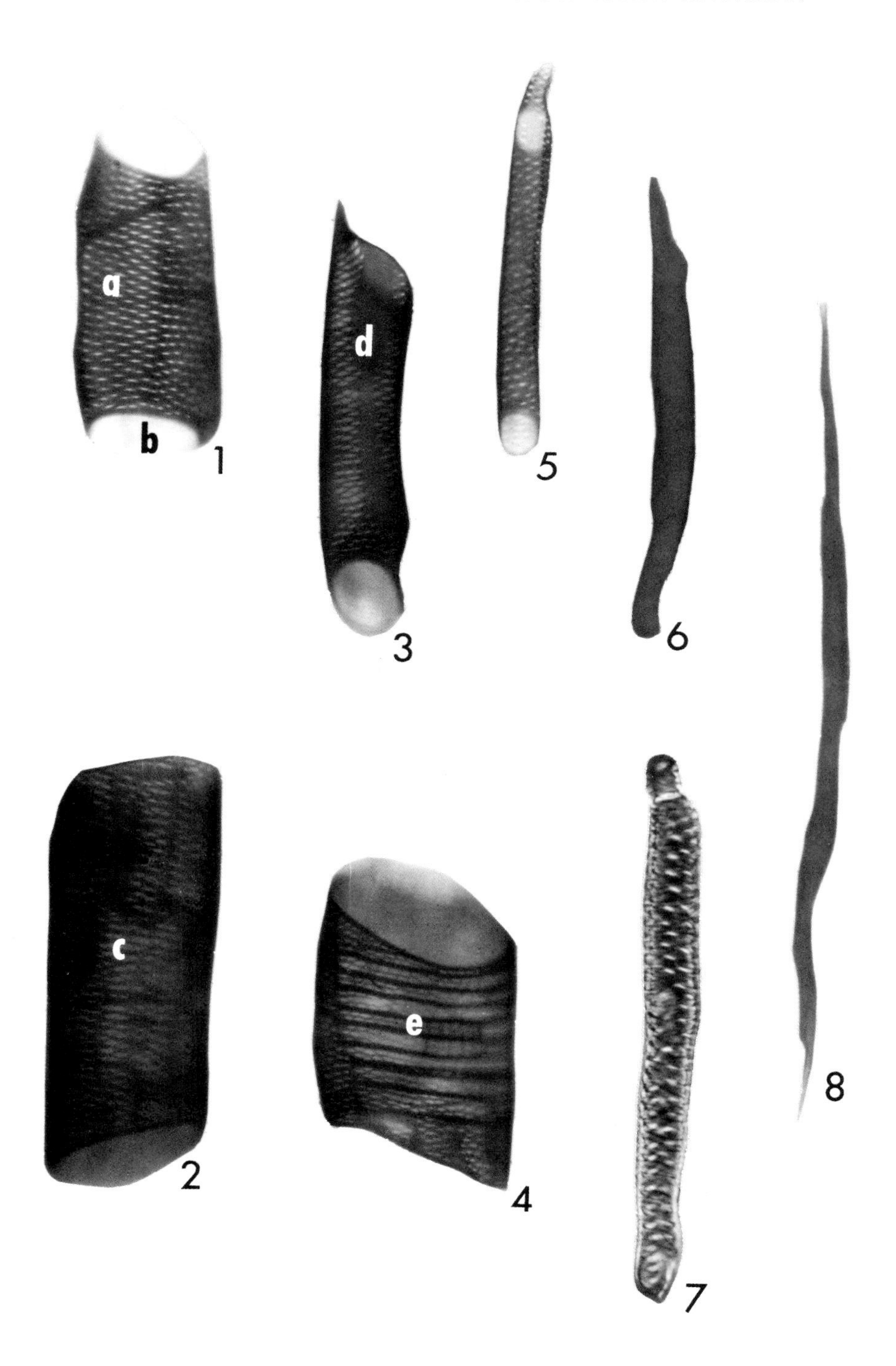

PLATE 64

CORN

Zea mays L.

1. Tracheid-like cell, 285X
2. Pitted vessel element, 185X
3. Portion of a vessel element, 185X, with annular thickenings
4. Parenchymatous cell, 185X, with simple pits
5. Vessel element, 185X, with scalariform-reticulate thickenings
6. Parenchymatous cell, 185X, from ground tissue
7. Ring-thickening, 185X, from an annular vessel element
8. Fiber, 185X
9. Tracheid, 185X
10. Short, wide-lumened fiber, 185X
11. Parenchymatous cell, 185X, with simple pits
12. Fiber, 185X

PLATE 65

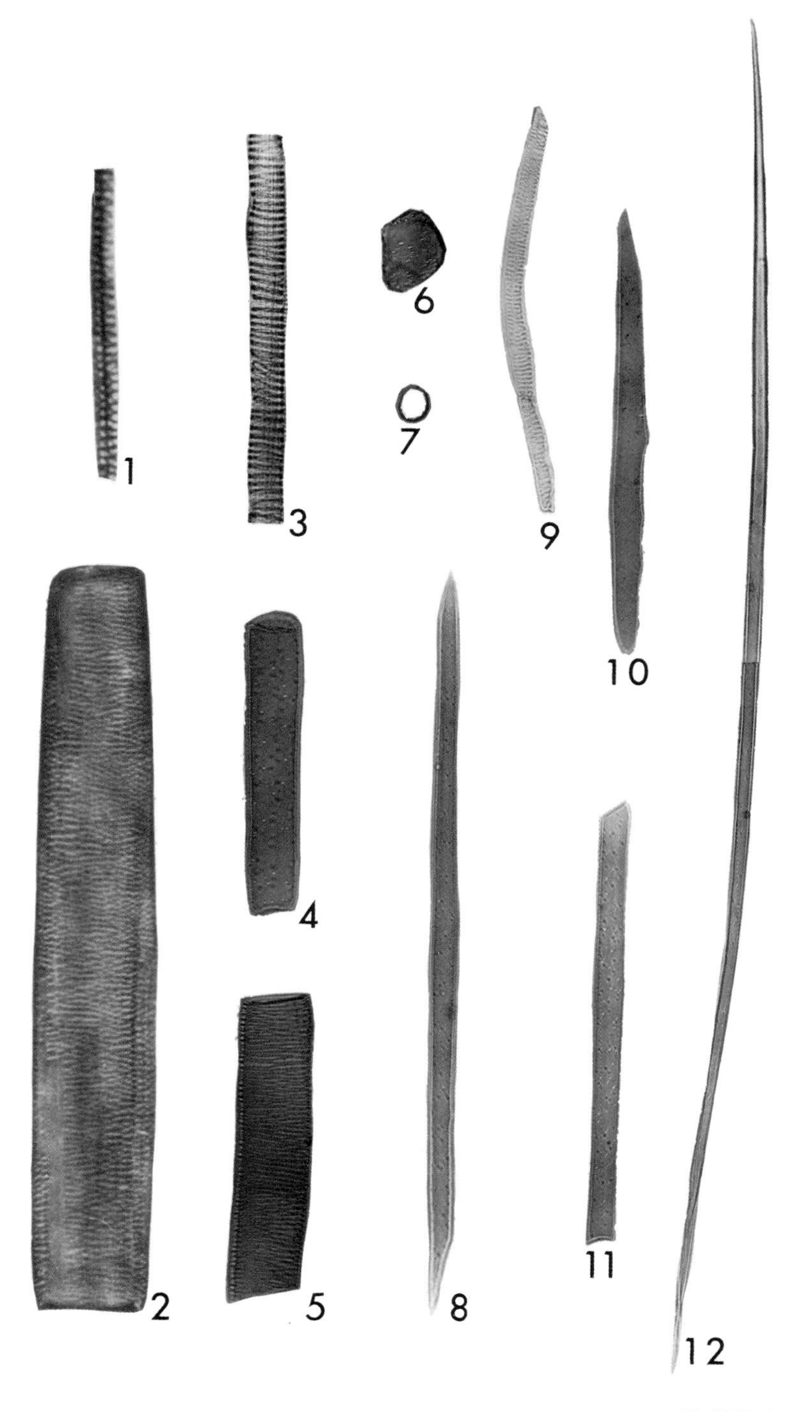

PLATE 65

SUGAR CANE
Saccharum officinarum L.

1. Parenchymatous cell, 185X
2. Pitted vessel element, 185X
3. Parenchymatous cell, 185X, with minute, simple pits
4. Pitted vessel element, 185X
5. Ring-thickening, 185X, from an annular vessel element
6. Several ring-thickenings, 185X, from an annular vessel element
7. Two parenchymatous cells, 185X, in contact lengthwise, as in the tissue
8. Wide-lumened fiber, 185X
9. Fiber, 185X

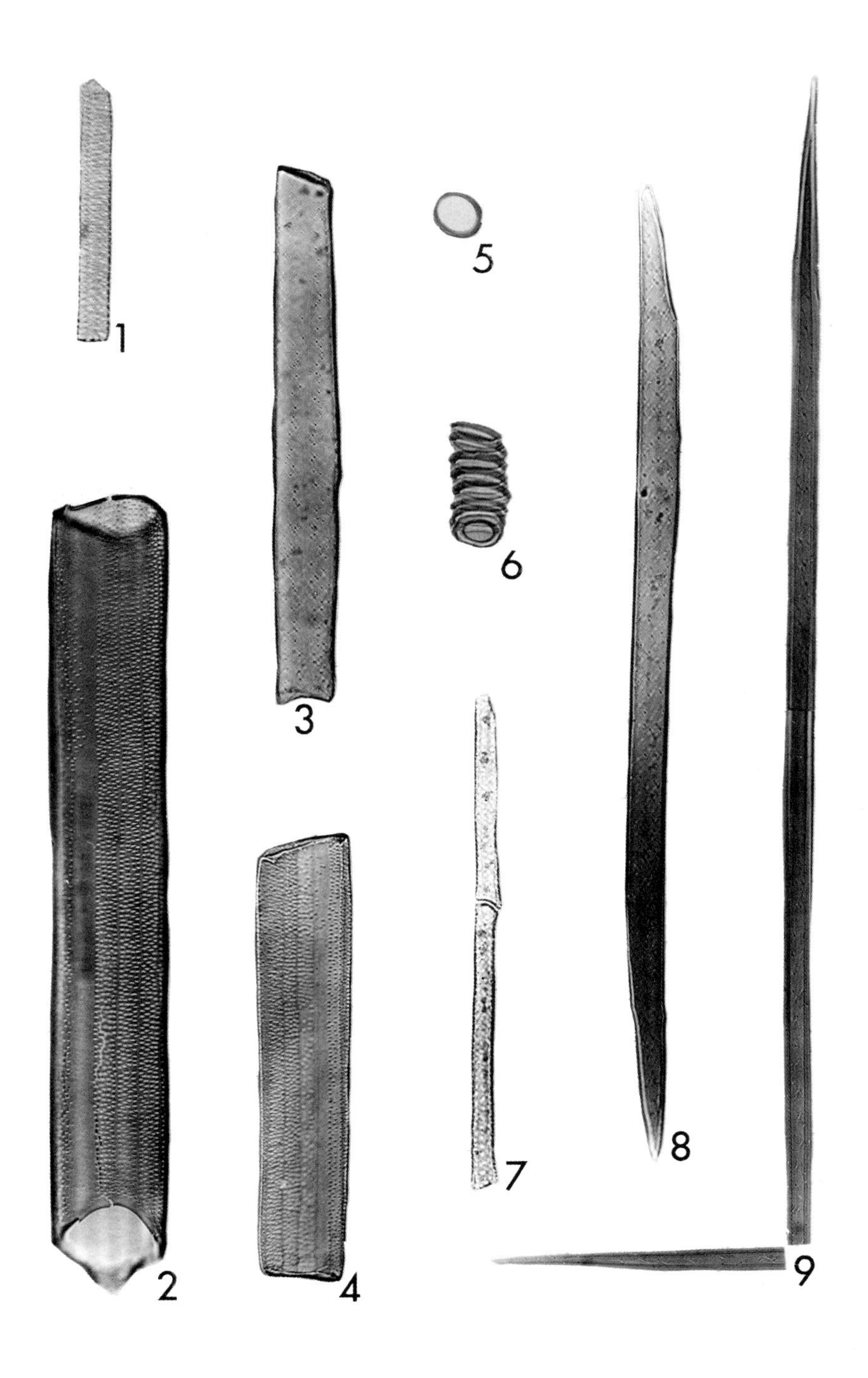

PLATE 66

BAMBOO

Cephalostachyum pergracile Munro

1. Short, tracheid-like cell, 285X
2. Pitted, short, tracheid-like cell, 285X, with helical thickening
3. Pitted vessel element, 95X, showing areas of contact with longitudinal parenchyma
4. Short fiber, 95X, notched at upper end
5. Long, parenchymatous cell, 95X
6. Cluster of parenchymatous cells, 95X, from ground tissue of the culm
7. Long, parenchymatous cell, 95X
8. Fiber, 95X
9. Fiber, 95X

PLATE 67

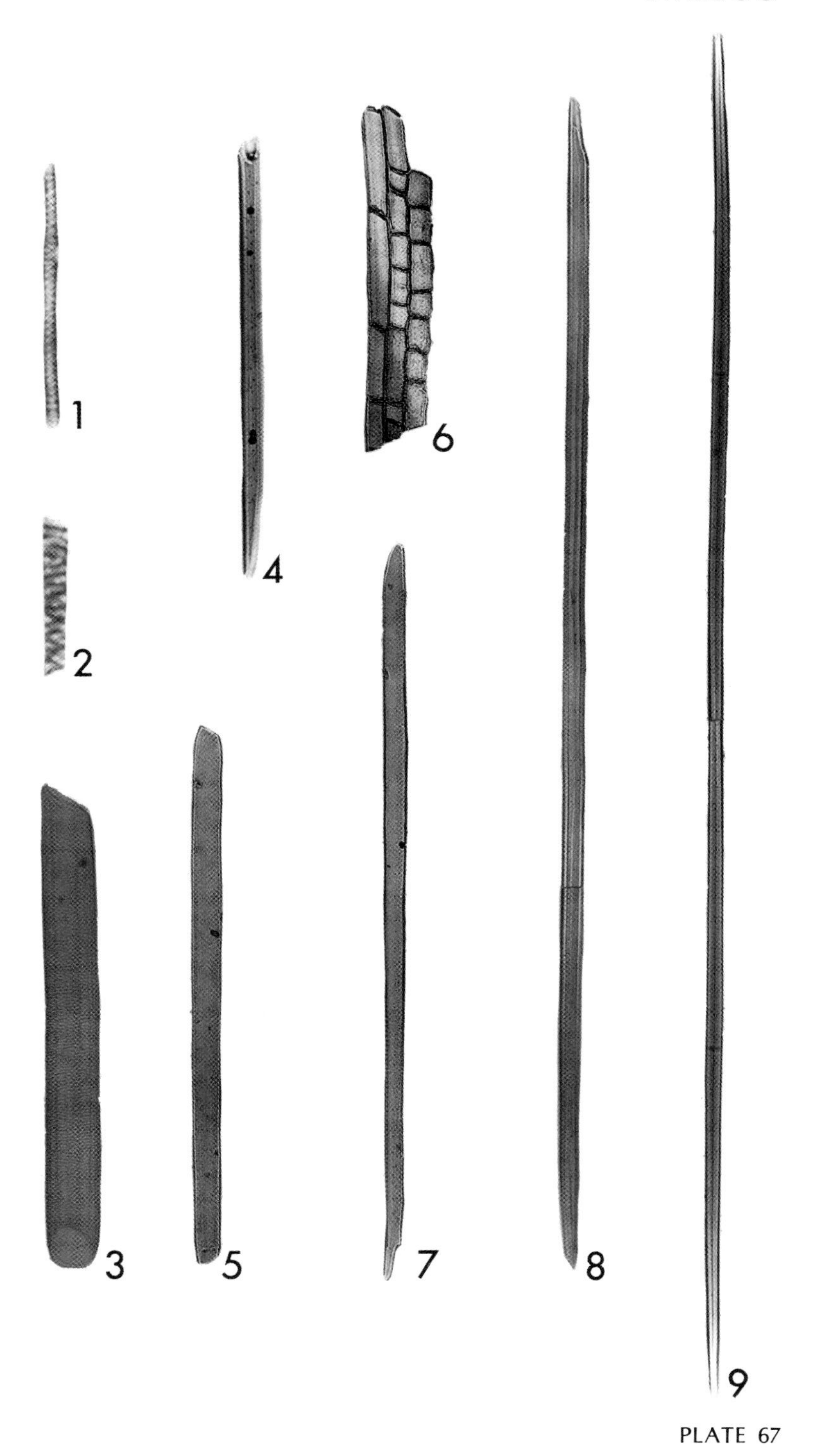

PLATE 67

WHEAT
Triticum sp.

1. Long, pitted vessel element, 185X
2. Narrow-lumened tracheid, 185X
3. Broad, parenchymatous cell, 185X
4. Portion of a vessel element, 185X, with annular thickenings
5. Parenchymatous cell, 185X, with sparse, simple pits
6. Long, parenchymatous cell, 185X
7. Fiber, 185X
8. Fiber, 185X

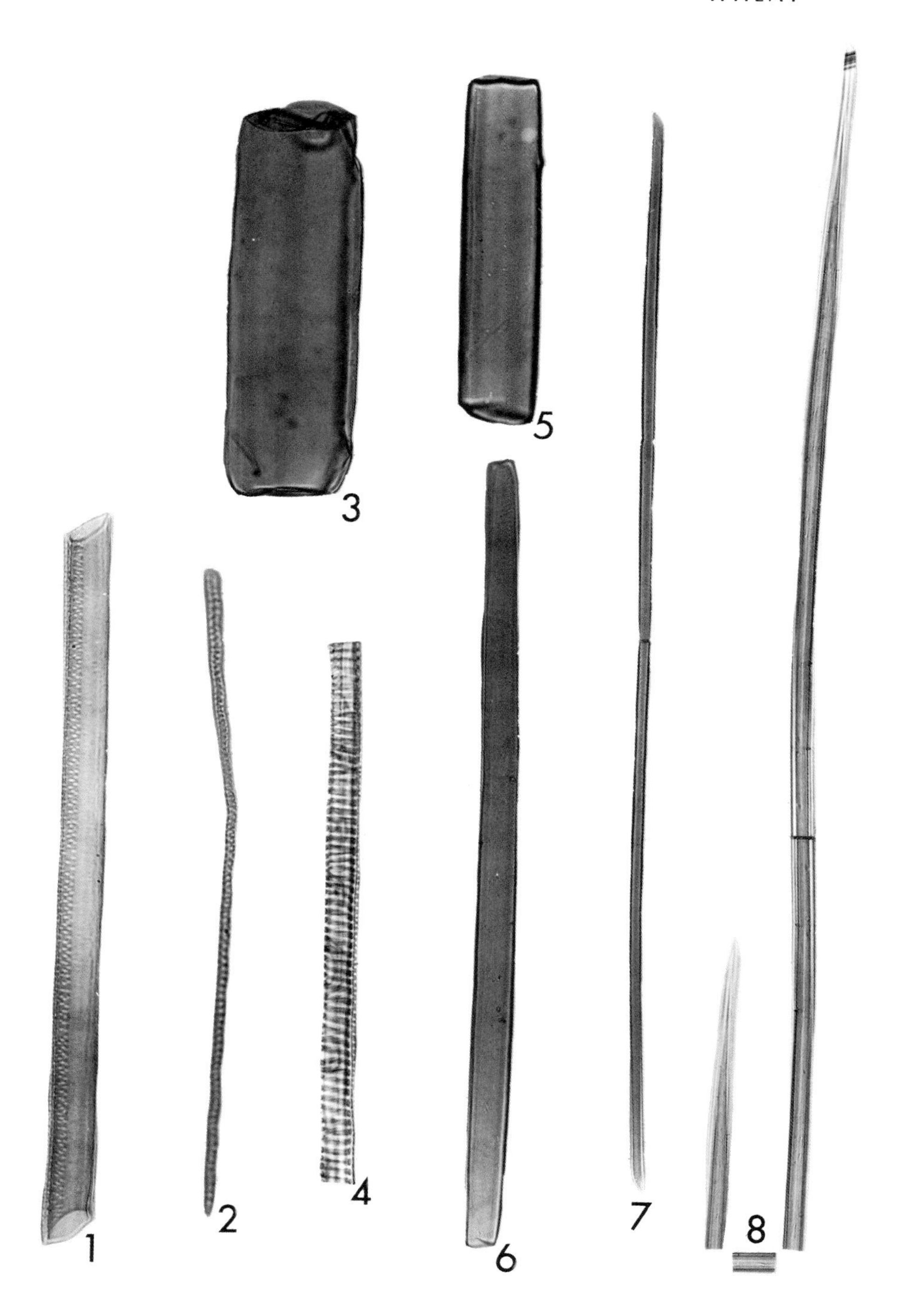

PLATE 68

RICE

Oryza sativa L.

1. Portion of leaf epidermis, 375X, surface view
   a. Stomate
   b. Epidermal cell with scalloped wall
2. Two parenchymatous cells, 375X, in contact lengthwise, as in the tissue
3. Pitted vessel element, 185X
4. Pitted vessel element, 185X, with two tracheids in contact, as in the tissue
5. Parenchymatous cell, 185X, from ground tissue
6. Tracheid, 185X
7. Parenchymatous cell, 185X
8. Parenchymatous cell, 185X
9. Long,narrow, parenchymatous cell, 185X
10. Fiber, 185X
11. Fiber, 15X

PLATE 69

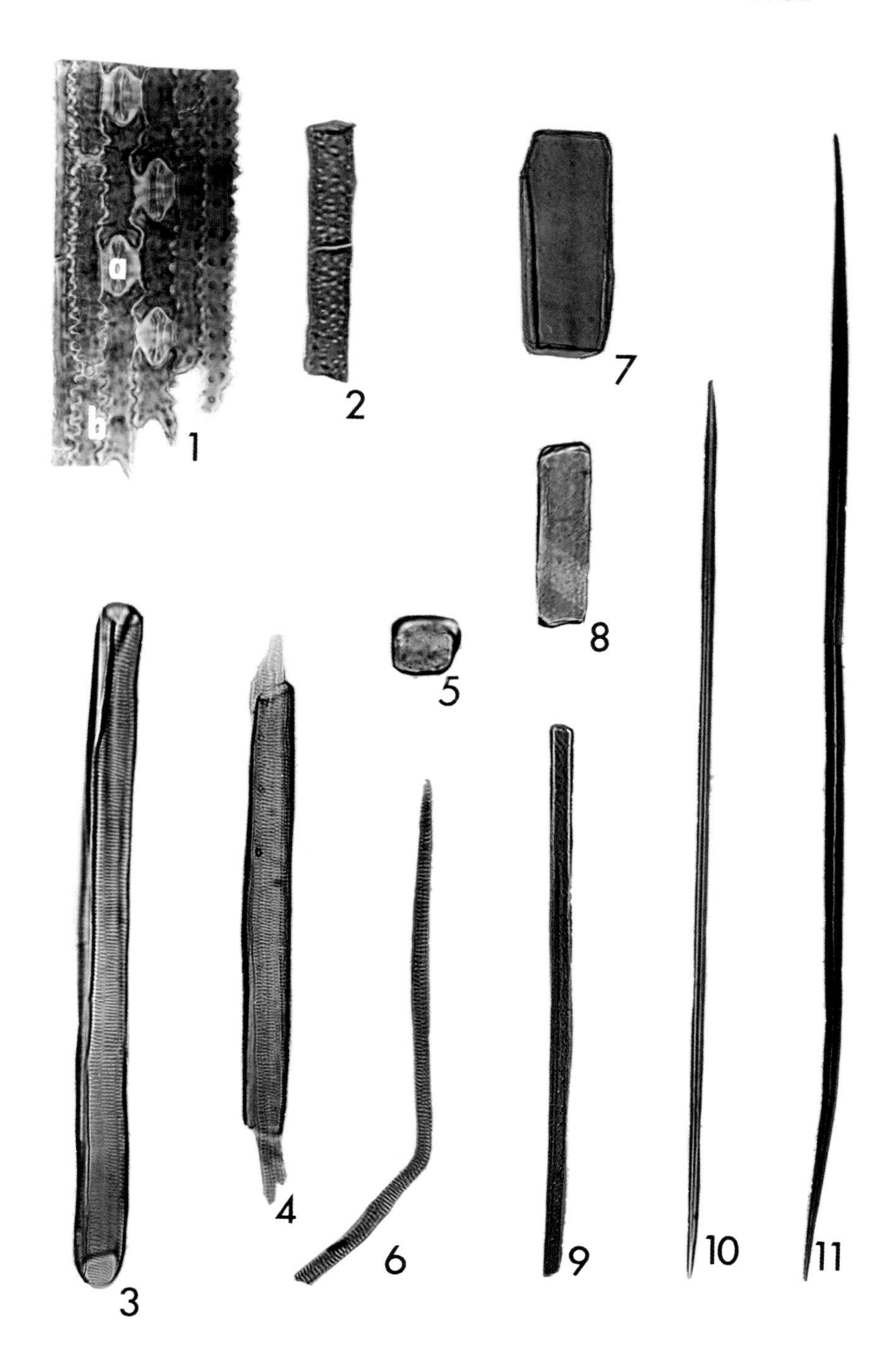

PLATE 69

ESPARTO GRASS
Stipa tenacissima L.

1. Portion of leaf epidermis, 375X, surface view
   a. Stomate
   b. Epidermal hairs seated on surface
   c. Epidermal cell with scalloped wall
2. Epidermal hair, 375X, lateral view, with epidermal cell attached, as in the tissue of the leaf
3. Pitted vessel element, 375X
4. Tracheid-like cell, 375X
5. Narrow tracheid, 375X
6. Series of irregular shaped, thick-walled cells, 375X, with intercellular spaces, lateral view. Cells of this type are present in the mixture of fibrous elements
7. Short, tracheid-like cell, 375X
8. Parenchymatous cell, 375X
9. Wide-lumened, fibrous cell, 375X
10. Fiber, 375X

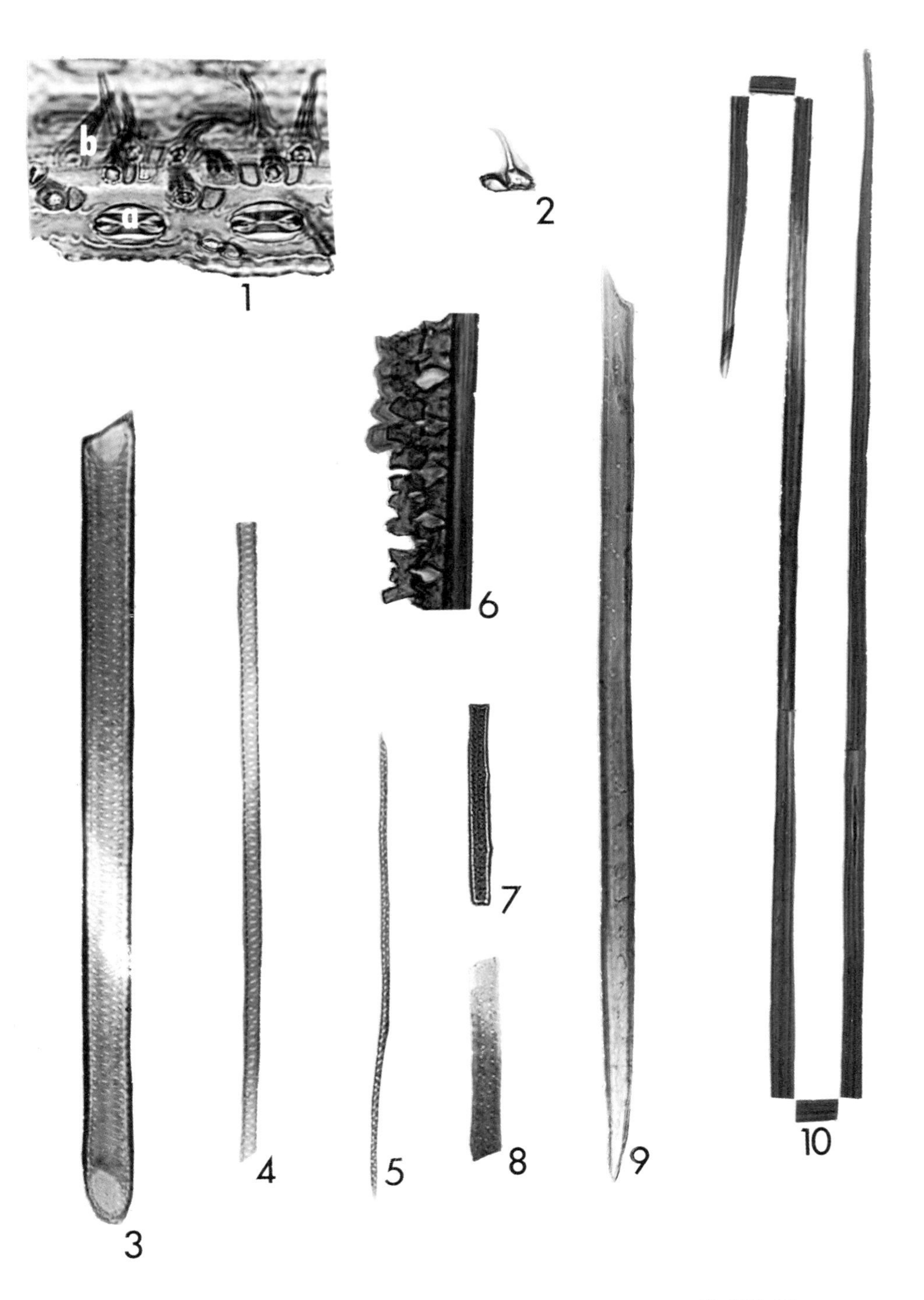
b
a
1
2
3
4
5
6
7
8
9
10

MANILA HEMP

Musa textilis Née

1. Portion of a leaf fiber, 185X
    a. Pits of pit pairs leading to a contiguous longitudinal fiber
2. Portion of a leaf fiber, 185X
3. Terminal portion of a leaf fiber, 185X

JUTE

Corchorus sp.

1. Portion of a bast fiber, 185X, showing variable width of cell lumen
2. Portion of a bast fiber, 185X
3. Terminal portion of a bast fiber, 185X

MANILA HEMP JUTE

HEMP

Cannabis sativa L.

1. Portion of a thick-walled bast fiber, 285X, with narrow lumen showing as a dark line
2. Portion of a thick-walled bast fiber, 285X
   - *a*. Failures in cell wall which develop during processing
3. Portion of a thick-walled bast fiber, 285X
   - *b*. Failures in cell wall which develop during processing

SISAL, HENEQUEN

Agave sp.

1. Terminal portion of a leaf fiber, 285X
2. Portion of a leaf fiber, 285X
3. Portion of a leaf fiber, 285X
   - *a*. Pits of pit pairs leading to a contiguous longitudinal fiber

HEMP

SISAL, HENEQUEN

PLATE 72

PAPER-MULBERRY
Broussonetia papyrifera Vent.

1. Central, broadened portion of a bast fiber, 285X, with outer lamella of wall showing as light-colored boundary layer

2. Portion near the end of a long bast fiber, 285X

PINEAPPLE
Ananas sp.

1. Portion of a leaf fiber, 375X

2. Portion of a leaf fiber, 375X

PAPER-MULBERRY

PINEAPPLE

RAMIE, CHINA GRASS

Boehmeria nivea (L.) Gaud.

1. Portion of a bast fiber, 185X
    - *a.* Nodelike failures in cell wall due to processing
    - *b.* Air bubble in lumen

2. Terminal portion of a bast fiber, 185X

3. Terminal portion of a bast fiber, 185X
    - *c.* Rounded tip

FLAX

Linum usitatissimum L.

1. Portion of a bast fiber, 185X, with a fairly wide lumen

2. Portion of a bast fiber, 185X, with a narrow lumen
    - *a.* Nodelike failures in cell wall due to processing

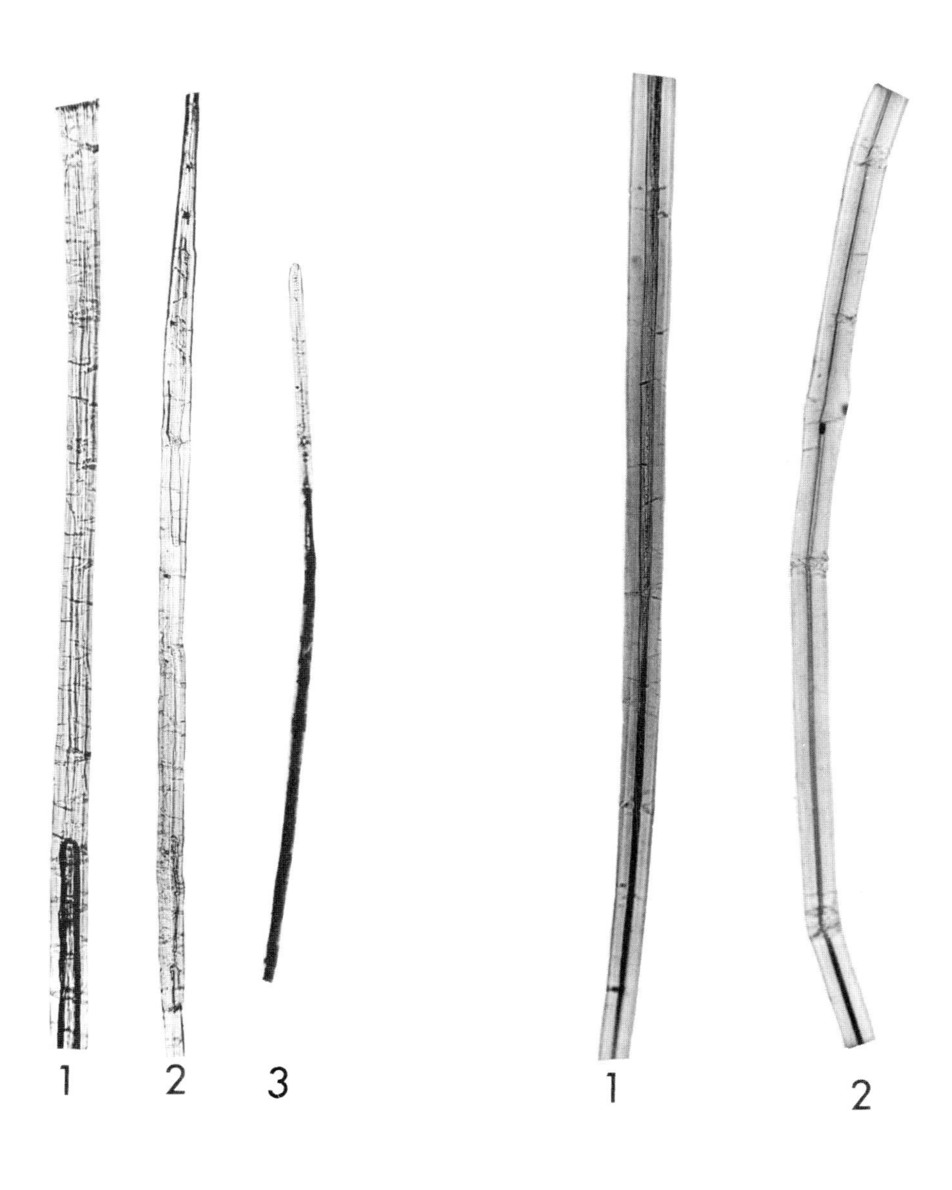

PLATE 74

COTTON

Gossypium sp.

1. Portion of a twisted seed hair, 185X
2. Portion of a twisted seed hair, 185X
3. Portion of a twisted seed hair, 185X
4. Cross-sections of seed hairs, 185X

KAPOK

Ceiba pentandra Gaertn.

1. Portion of a thin-walled, wide-lumened seed hair, 185X
2. Portion of a thin-walled, wide-lumened seed hair, 185X

PLATE 75

COTTON

KAPOK

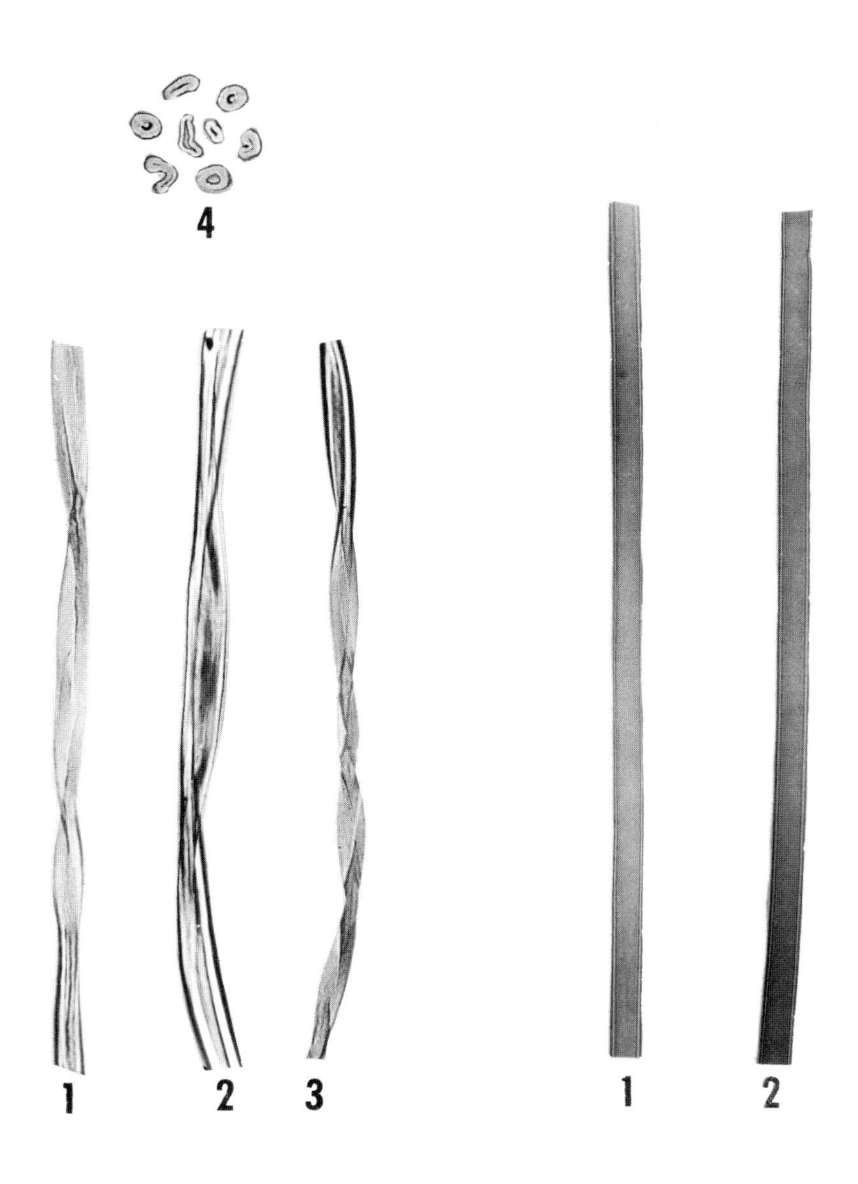

WOOL

1. Portion of a hair, 185X
    - *a.* Imbricated scale cells
    - *b.* Medulla

2. Portion of a hair, 185X

SILK

1. Portion of a silk thread, 185X, consisting of several double strands
2. Portion of a degummed strand, 185X
    - *a.* Indication of double nature of strand
3. Portion of a degummed strand, 185X

PLATE 76

WOOL

SILK

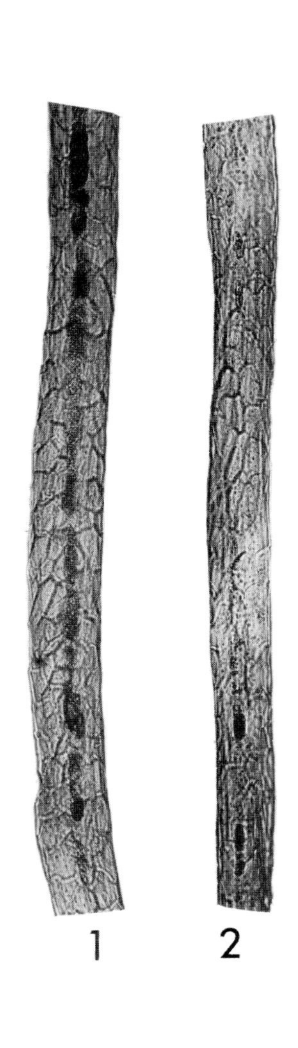

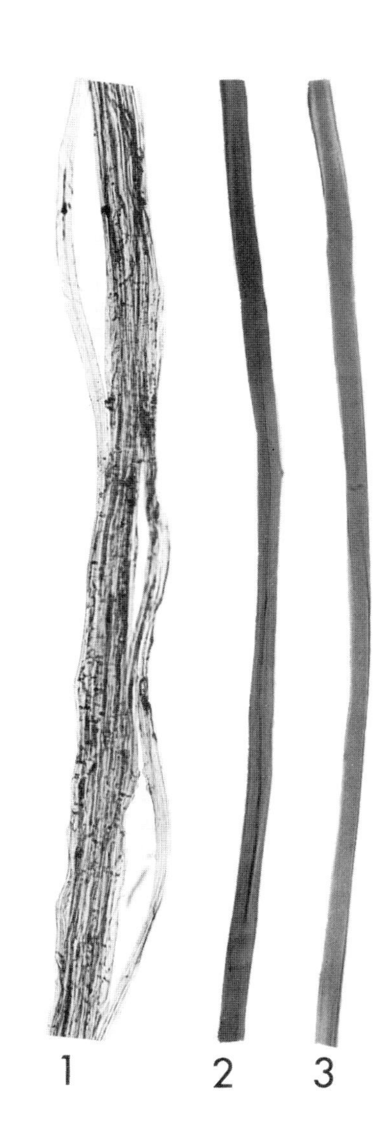

PLATE 76

ASBESTOS

Sample of a commercial asbestos paper, 185X

a. Aggregate consisting of thin, filamentous crystals

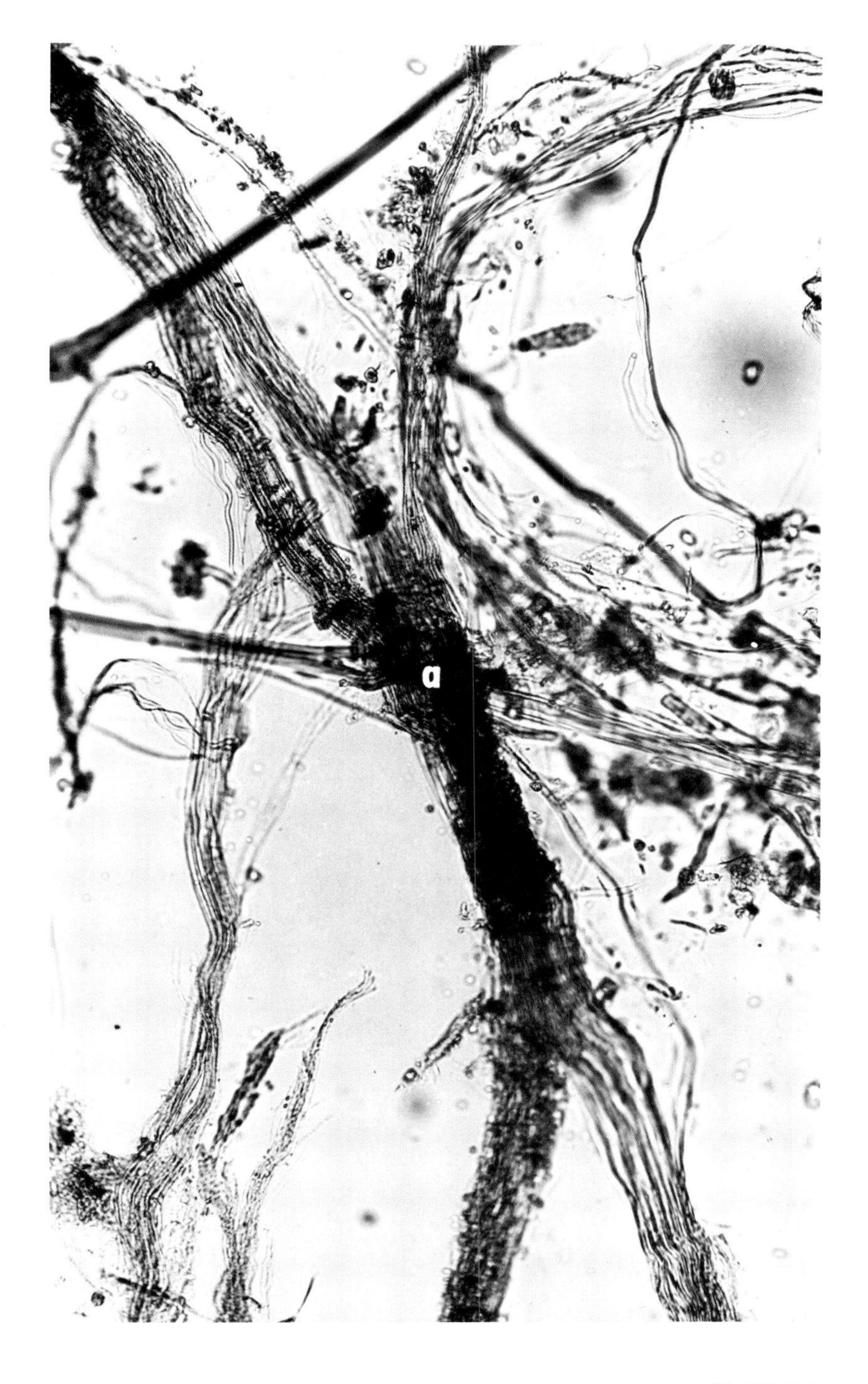
a

RAYON

1. Portion of a fiber, 375X
2. Portion of a fiber, 375X

NYLON
(Containing a delustrant)

1. Portion of a fiber, 375X
2. Portion of a fiber, 375X

PLATE 78

RAYON NYLON

PLATE 78

MAN-MADE FIBERS

Cross-section, 285X

1. Cellulosic fiber. Regular rayon
2. Cellulosic fiber. High tensile rayon. "Avril"
3. Polyester fiber. "Dacron."*
4. Cellulosic fiber. XL HiStrength rayon
5. Cellulosic fiber. Secondary acetate
6. Acrylic fiber. "Orlon"*
7. Polyamide fiber. Nylon

*Registered Du Pont Trademarks.

Acknowledgment is made to American Viscose Corporation for providing examples of cellulosic fibers shown here and to E. I. Du Pont de Nemours & Company for supplying samples of "Orlon,"* "Dacron"* and nylon.

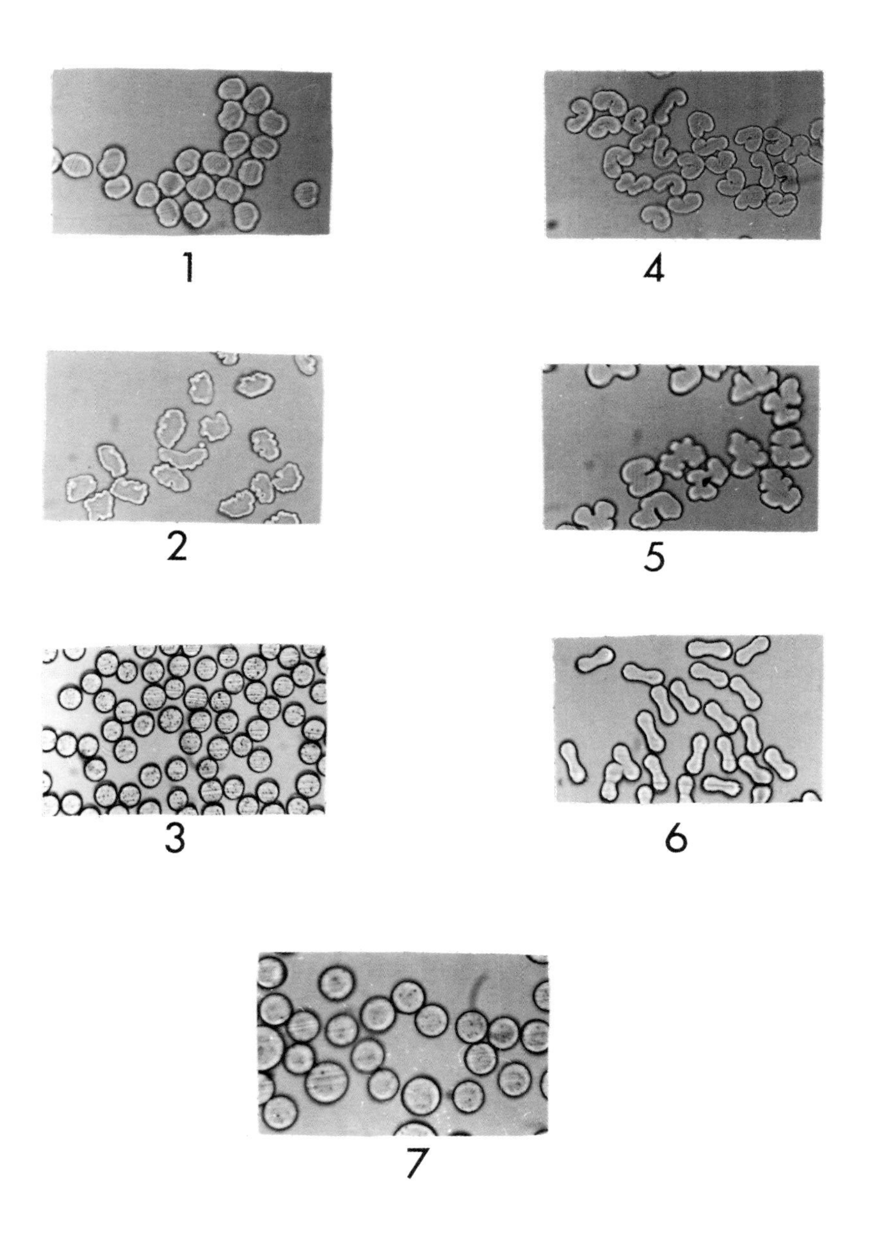
1
4
2
5
3
6
7